International School of Cosmology and
Gravitation XVI Course

CLASSICAL AND QUANTUM NONLOCALITY

THE SCIENCE AND CULTURE SERIES — PHYSICS

Series Editor: A. Zichichi, *European Physical Society, Geneva, Switzerland*

Series Editorial Board: P. G. Bergmann, J. Collinge, V. Hughes, N. Kurti, T. D. Lee, K. M. B. Siegbahn, G. 't Hooft, P. Toubert, E. Velikhov, G. Veneziano, G. Zhou

1. Perspectives for New Detectors in Future Supercolliders, 1991
2. Data Structures for Particle Physics Experiments: Evolution or Revolution?, 1991
3. Image Processing for Future High-Energy Physics Detectors, 1992
4. GaAs Detectors and Electronics for High-Energy Physics, 1992
5. Supercolliders and Superdetectors, 1993
6. Properties of SUSY Particles, 1993
7. From Superstrings to Supergravity, 1994
8. Probing the Nuclear Paradigm with Heavy Ion Reactions, 1994
9. Quantum-Like Models and Coherent Effects, 1995
10. Quantum Gravity, 1996
11. Crystalline Beams and Related Issues, 1996
12. The Spin Structure of the Nucleon, 1997
13. Hadron Colliders at the Highest Energy and Luminosity, 1998
14. Universality Features in Multihadron Production and the Leading Effect, 1998
15. Exotic Nuclei, 1998
16. Spin in Gravity: Is It Possible to Give an Experimental Basis to Torsion?, 1998
17. New Detectors, 1999
18. Classical and Quantum Nonlocality, 2000

THE SCIENCE AND CULTURE SERIES — PHYSICS

International School of Cosmology and
Gravitation XVI Course

CLASSICAL AND QUANTUM NONLOCALITY

Erice, Italy 27 April – 4 May 1999

Editors

P. G. Bergmann
New York University

V. de Sabbata
University of Bologna

J. N. Goldberg
Syracuse University

Series Editor
A. Zichichi

World Scientific
Singapore • New Jersey • London • Hong Kong

Published by

World Scientific Publishing Co. Pte. Ltd.
5 Toh Tuck Link, Singapore 596224
USA office: 27 Warren Street, Suite 401-402, Hackensack, NJ 07601
UK office: 57 Shelton Street, Covent Garden, London WC2H 9HE

British Library Cataloguing-in-Publication Data
A catalogue record for this book is available from the British Library.

The Science and Culture Series - Physics
CLASSICAL AND QUANTUM NONLOCALITY

ISBN-13 978-981-02-4296-1
ISBN-10 981-02-4296-4

PREFACE

This Course has provided an up-to-date understanding of the progress and current problems of interplay of non-locality in classical theories of gravitation and quantum theory. These problems are on the border area between general relativity and quantum physics including quantum gravity. Classically non-locality appears whenever one considers a curved space-time.

According to general relativity the field of gravitation is identical with the geometric structure of space-time. Only when gravity is extremely weak does it make sense to split that into a "background metric" (a Minkowski space-time) and a dynamic field of variables, which in first approximation obey linear laws. As mankind approaches the age of deep exploration, studies of strong gravitational fields assume increasing relevance.

An important aspect of non-locality rests on the fundamental arbitrariness of space-time coordinates. No matter how completely Cauchy data are prescribed on a three-dimensional (hyper-) surface, propagation of that surface is not unique. The generator of such a propagation is an integral over the surface. Its integrand is a linear combimation of four constraints, each multiplied by a highly arbitrary factor, which in general involves the canonical field variables.

Whenever the curvature of curved space-time is important, the localization of the events involves serious observational and conceptual ambiguities. Therefore physical quantities may lose their local meaning. This is particularly true for the notion of energy. However in quantum theory non-locality of events is fundamental. The concept of non-locality appears at the very foundation of quantum theory in the famous debate between Einstein and Bohr culminating in the Einstein-Podolski Rosen-argument.

These ambiguities assume critical importance near the Planck length $(\hbar G/c^3)^{1/2} \approx 10^{-33}$cm, at which the Schwarzschild radius and the Compton wavelength of a massive particle are of the same order of magnitude. Though the Planck length is small compared the lengths encountered in particle physics, the corresponding Planck mass, $\approx 10^{-5}$g, is huge.

Non-locality enters space-time not only through the Planck length, but as a concomitant of Heisenberg's uncertainty relations. As geometry and dynamics cannot be separated, distances can be "sharp" only if momentum densities are indefinite. As a result the existence of world points apart from the fields defined on them may well be a concepts which needs modification in the classical (non-quantum) as well as in quantum versions of the gravitational physics.

All these facets of non-locality are characteristic not only of Einstein's general relativity, but of any theory whose field equations are derived from an action integral that is form-invariant under arbitrary curvilinear coordinate transformations ("diffeomorphisms"). Formally such general relativistic theories differ from Poincaré-invariant theories with (non-abelian) gauge variables in that the latters' invariance group is homomorphic to the Poincaré group, whereas in general-relativistic theories there is but one invariance group, free of any normal subgroups. Because of this difference in the structures of the respective invariance groups we may need to change some basic concepts, that properly belong to special relativity.

Past Courses arranged by the International School of Cosmology and Gravitation have been devoted to diverse aspects of gravitation. This XVI Course explored the role of the space-time point in the conventional space-time manifold and its possible modifications, ranging from the Planck length to the Universe.

In the lecture articles one can find an outline of the epistemological background of the Einstein-Bohr debate and also the Bohr's and Mach's conception of non-locality in gravitation; a description of global energy and non-local energy with the concept of quasi-local energy; different aspects of physical nonlocality and some problems of nonlocality in quantum physics, the EPR argument and quantum non-locality, different conceptions of non commuting geometry, some topics and trials on the theory of non-local fields together a theory of fields in Finsler space and fluctuation of space-time, the Planck-scale physics and non-locality (including black holes), the nonlocality connected with Bell-inequality experiments and quantum state teleportation, the search for quantum gravity with the latest developments in cosmology, and other arguments connected with non-locality.

The Editors wish to conclude by thanking the Ettore Majorana Center for Scientific Culture which made a considerable financial contribution for the Course. We should express our great appreciation for the yeoman's service rendered by Dr. Alberto Gabriele who, in fact, compensated the big vacuum created by the enforced absence of all the Directors of the School due to their illness. We are very much shocked for and deeply regret the sudden demise of Alberto Gabriele while at work in Erice on the evening of 27th June, 1999. For thirthy-five years Dr.Gabriele has carry out the difficult task of running every School and solving the accompanying problems with exceptional intelligence, great patience and human feelings.

We should be thankful to Prof.Arthur Komar who conducted the day-to-day academic programmes of the Course very actively by his great erudition. In fact Professor Komar excellently did the job of the Directors of the School who failed to attend the Course due to their illness.

The Editors

Peter G.Bergmann
Venzo de Sabbata
Joshua Goldberg

CONTENTS

BOHR'S AND MACH'S CONCEPTIONS OF NON-LOCALITY IN GRAVITATION

HORST-HEINO V. BORZESZKOWSKI
Technical University Berlin, Institute for Theoretical Physics, Hardenbergstrasse 36, D-10623 Berlin; e-mail: borzeszk@itp.physik.tu-berlin.de; Germany

and

HANS-JÜRGEN TREDER
Rosa-Luxemburg-Str. 17a, D-14482 Potsdam; Germany

ABSTRACT

Local general relativity theory and non-local Mach-Einstein mechanics are extremely opposite attempts to realise the equivalence of inertia and gravitation. In the first case one is missing certain telescopic elements in the sense of Mach's principle, in the second one arrives at a mechanics that can hardly recover the results of the successful local physics. Therefore, in order to introduce telescopic elements into gravitational theory, one should choose a way between these two extremes. – It will be shown that, in the framework of tetrad theories of gravitation stemming from a Lagrangian built from Weitzenboeck's (anholonomic) invariants, one can harmonise Mach's principle with the locality principle by restricting general relativity. These theories satisfy a (local) field theoretical version of the Mach-Einstein doctrine saying that, via gravitation, the motion of matter described in concordance with the principle of equivalence determines the reference systems. It holds an interesting "complementarity" between strong general relativity of GRT and Mach-Einstein doctrine. In GRT, via Einstein's equations, the covariant and Lorentz-invariant Riemann-Einstein structure of the space-time defines the dynamics of matter. Otherwise, in all other cases, i.e., for the theories here under consideration the matter determines, via gravitation, the reference systems up to global Lorentz transformations.

1. Locality vs. Non-Locality: On microscopic and telescopic approaches to gravitation

Max Planck [60] pointed out that the local and global (non-local) points of view are not equivalent:[1] According to the field-theoretical locality principle, the forces acting at a point depend only on the infinitesimal vicinity of this point (therefore, Planck called a field theory an "infinitesimal theory"), while any non-local action-at-a-distance theory holds that these forces are determined by all bodies of the universe. Accordingly, Planck concluded that the locality principle, postulating the reducibility of all non-local interactions ("telescopic" interactions called by him) implies a simplification of, and also a restriction upon, the nature and mode of action of all natural forces. For him, the answer to

[1] For the following passages, see also Ref. 70, p. 109 *fortissimo*.

the question, whether the non-local (telescopic) action-at-a distance or local (infinitesimal, or, as called by C. Neumann, "microscopic") field theory is valid, depended on the measure of success achieved by the respective theory in describing interactions.[2]

As to gravitation based on the principle of equivalence, this means that one can raise the question whether the latter principle should be realised microscopically or telescopically. Einstein's GRT realises it microscopically by reducing gravitation to inertia, i.e., to free motion in a metric field. In this theory, the one-body problem is the appropriately posed problem; it leads, in the case of monopole test particles, to geodesic motion. In its quantised version, this local theory leads inevitably to planckions which indicate that, according to a common expectation, the standard picture of space and time is applicable only at scales well above the Planck regime [50].[3] Planckions appear in quantum GRT as an implication of Bohr's non-locality condition $\Delta x \geq c\Delta t$ to be imposed on the extent of the measurement region and the duration of the measurement procedure (Bohr & Rosenfeld [12]). They replace the space-time points and the point particles of field theory and impose this way a limitation on special-relativistic quantum field theory. In this sense planckions signal the existence of limitations of the local description. This might be considered as one reason to turn to a telescopic realisation of the equivalence principle. Further reasons are:[4] (1) Despite continued success of field theories, theoreticians have not succeeded in eliminating all telescopic elements from local, i.e., field, theories. Prominent among these elements are the initial and boundary conditions, which are necessary for the integration of the field equations. (Therefore, local physics must be supplemented by telescopic elements.) (2) The power of the local-theory conception derives possibly from the nature of the interaction. And, since gravitation is the most "telescopic" interaction we know, a breakdown of the local description of interactions is most likely to occur in the area of gravitational interaction. (According to our above-given argument, the latter seems to be corroborated in the quantum regime, where gravitation leads even to limitations on relativistic field theory.)

Now, turning to a telescopic description, let us first ask what telescopic elements one can assume to start with. As to this question, it is helpful to remind of the following point.[5] Due to the telescopic elements missing in local theories, the relativistic field theory realises its principles of relativity and covariance by offering an infinite variety of "universes" as mathematically possible and virtually contained in the variety of possible world models. However, the actual cosmos specifies one and only one reference system as its "proper system".

Indeed, in the actual cosmos investigated by astronomy there is an empirical equivalence of three reference systems which are introduced by three different methods:

---

[2] Moreover, according to Planck, the global aspect is always the more general one, in the sense "that a finite quantity includes an infinitesimal as a special case."

[3] To our mind, at distances of the order of magnitude of Planck's length, the quantum fluctuations induced by the measuring procedure are so violent that this regime is not accessible to measurement (see Ref. 15) This interpretation is especially plausible when one follows Rosenfeld's measurement-theoretical considerations (for this, see the last section).

[4] Cf. Ref. 70, p. 110.

[5] Cf. Ref. 69, p. 46 *fortissimo*.

The inertial system of celestial mechanics determined dynamically, the astronomical fundamental system given by the galaxies, and the black background radiation defining the same class of reference systems.[6] This coincidence seems to imply a cancellation of the relativity principles.

From the viewpoint of a field theory these reference systems defined macroscopically are equivalent to a purely accidental choice of boundary and initial conditions. The cosmological world models of GRT specifying these reference systems are, from the viewpoint of GRT, certain special solutions of Einstein's equations with matter. However, the cosmos is unique (at any rate, there is no other assumption which is epistemologically satisfying). With respect to this assumption, on the one hand, the microscopic principles do too much; they provide not only the actual cosmos but an infinite variety of models, as well. On the other hand, they are not able to determine macroscopic physics at all: They furnish only necessary but no sufficient conditions. The microscopic character of field physics imposes a special supposition on the physical laws, in particular, leading to the fact that they have to be formulated as differential equations. Therefore, local physics must be supplemented by telescopic (integral) principles: In general integral principles uniquely lead to differential equations, while the inverse operation is ambiguous. (This reminds of Planck's statement mentioned above.)

The concept of an absolute space determined by the cosmic masses (let us call this conception "Mach principle") implies such telescopic elements. However, this space cannot be the Newtonian space since, despite Mach's attempts of reinterpreting Newtonian mechanics, it is not determined by the cosmic masses [and it is the proximity of GRT to Newton's theory that it makes so difficult to realise Mach's principle in the GRT framework (Ref. 70)].

However, the problem is that the telescopic theory (in the sense of Planck) cannot be formulated either in the three-dimensional space or in the four-dimensional space of the strictly local theory of relativity. From the telescopic point, every degree of freedom of a physical system provides generally one dimension of the configuration space V_{3N}.

Assuming, according to H. Hertz [49], a universe of N atomic particles of equal mass represented by mass points, the space to be used in telescopic physics is given by the $3N$-dimensional configuration space V_{3N} considered by Hertz. In general, there exist connections and interactions which are anholonomic, i.e., nonintergrable. A general and complete separation of the Hertzian $3N$-dimensional space of the single particles A,

$$V_{3N} = \prod_{A}^{N} V_3^{(N)},$$

is generally only possible if the connections between the particles (i.e., the constraints) are holonomic and accordingly capable of being representing the form of Newtonian forces.

If the particles are identical, this is assumed in classical physics especially by Hertz, then the N factor spaces V_3 may also be identified. Thus, for holonomic connections, the three-dimensional absolute Newtonian space is again obtained. There is an obvious analogy to Schrödinger's wave mechanics: It is also formulated in a configuration

[6] The isotropy of the background radiation even leads to Newton's absolute reference system.

space. But, unlike our mechanics, in quantum mechanics the particles are not identical, such that the partial spaces may not be identified. (Just this is reflected by the Einstein-Podolsky-Rosen paradox.) In the Hertzian configuration space, for the "total universe" it is now possible to formulate the dynamics with the respect to the cosmic particles themselves and, accordingly, to use in dynamics only reciprocal functions of the particles, as was required by Huygens, Leibniz, and Poincaré. Following Mach and Einstein, then one has to postulate that the inertia of the N point masses A is completely induced by their mutual gravitation; inertia has to be described by a homogeneous scalar function of the local gravitational potential. Its main part stems from the average gravitational potential of the universe.

In contrast to GRT, now the equivalence of inertia and gravitation means a reduction of inertia to the quasi-Newtonian gravitational interaction. The metric of the Hertzian configuration space V_{3N} of the N particles is also a homogeneous function of the gravitational interaction between these particles. This principle combining the equivalence between inertia and gravitation with Mach's principle of the relativity of inertia (and thus with the postulate that there are cosmically determined reference systems) we call "Mach-Einstein doctrine".[7] The reference to the totality of cosmic masses implies a basically non-local approach to gravitation.

If one assumes that the Machian cosmos consists of N particles of the same (heavy) mass m, then, due to the Mach-Einstein doctrine (in its "quantised" version), the Bohr radius of two particles which are purely gravitationally coupled is given by the radius of the universe (the following relations (1)-(6) are order-of-magnitude-relations),

$$\frac{GM}{c^2} \equiv \frac{\hbar^2}{Gm^3} \qquad \text{(where } M = Nm \text{ denotes the total mass),} \tag{1}$$

so that one obtains

$$\frac{GM^2}{\hbar c} = \left(\frac{\hbar c}{Gm^2}\right)^3 \qquad \text{or} \qquad M = Nm = \left(\frac{\hbar c}{Gm^2}\right)^2 m \tag{2}$$

where

$$N = \left(\frac{\hbar c}{Gm^2}\right)^2 \tag{3}$$

is Eddington's number of particles with mass m which in the case of baryons amounts to $\omega = 10^{79}$.

Otherwise, considering the smallest Machian cosmos consisting of two particles of mass $\overline{m}$ one has to start with the relation

$$\frac{G\overline{m}}{c^2} \equiv \frac{\hbar^2}{G\overline{m}^3}. \tag{4}$$

Then one obtains for the mass of the particles the formula

$$\overline{m} = \left(\frac{\hbar c}{G}\right)^{1/2} \tag{5}$$

and, furthermore, the relations

$$\frac{G\overline{m}}{c^2} = \frac{\hbar}{\overline{m}c} = \left(\frac{\hbar G}{c^3}\right)^{1/2}, \quad \frac{G\overline{m}^2}{\hbar c} = 1, \quad \text{and} \quad \overline{m} = \left(\frac{\hbar c}{Gm^2}\right)^{1/2} m. \tag{6}$$

---

[7] For details, cf. Ref. 70 and the literature cited therein.

That means, in this case one is led to planckions such that planckions do not only appear in gravito-dynamics as smallest test body and biggest elementary particle, but also as a self-consistent object in Machian dynamics, namely as the smallest possible cosmos.

In a Machian "quantum gravito-dynamics", the fundamental set of constants $\hbar$, G, c is supplemented by the total gravitational mass M of the universe to which this dynamics is referred. As it can be seen from (1) and (2), the mass of the particles is determined by the cosmos, where the ratio $\hbar c/Gm^2$ is a dimensional constant (in the sense of Einstein's idea of an arithmetisation of physics).

2. Einstein's early attempt to realise telescopic elements in a relativistic field theory

The local approach of GRT reducing gravitation to inertia and the non-local approach based on the Mach-Einstein doctrine and reducing inertia to gravitation are extremely opposite attempts to realise the equivalence of inertia and gravitation. In the first case one is missing certain telescopic elements, in the second one arrives at a mechanics that can hardly recover the results of the rather successful microscopic (or local) physics. It seems to be evident that, in order to harmonise Mach's principle with general relativity, one has to choose a way between these two extremes.

To our mind, it was this harmonisation of local and nonlocal aspects which Einstein was looking for in the years 1913/14 and which led him to general relativity on a winding road. In particular, in reading Einstein's papers and notes made for himself in their context one sees that the *Ansätze* of 1913 and 1914 which, estimated from the viewpoint of the final formulation of GRT, appear as mathematically still incorrect had also a physical motivation which does not allow one to call them mathematically false. The material now generally accessible (see Refs. 25 and 26) makes it obviously that Einstein's doubts about the form of the relativistic gravitational equations was also due to the fact that the physical content of the principle of general relativity was not clear: Because, for Einstein, Mach's principle was an essential aspect of general relativity at that time he was even ready to accept restrictions on general covariance and thus of general relativity when this would make the gravitational theory "Machian".

In 1913 and 1914 Einstein [27, 28, 31, 32, 33] (partly in joint papers with Grossmann) argued as follows: Matter described by its symmetric energy-momentum tensor T_{ik} determines the metrical tensor g_{ik} of a Riemannian space-time up to affine.(i.e., linear) transformations

$$\hat{x}^i(x^l) = a^i{}_k x^k + b^i, \qquad \text{where } a^i{}_{k,l} = 0, \quad b^i{}_{,l} = 0. \tag{7}$$

The preferred coordinate systems related by Eqs. (7) are specified by the covariantly written dynamical equations

$$\left(\sqrt{-g}T^k{}_i\right)_{;k} = \left(\sqrt{-g}T^k{}_i + \sqrt{-g}t^k{}_i\right)_{,k} = 0. \tag{8}$$

These equations are the special-relativistic laws of energy-momentum conservation which in the case where a gravitational field is present and under the condition that Einstein's principle of equivalence is satisfied, necessarily, must be written in this covariant form.

As was stated by Einstein (1914),[8] this specification of the physically preferred reference systems is an expression of Mach's principle because it is an implication of the motion of (cosmic) matter described by Eqs. (8). This principle determines those reference systems which should replace the Newtonian inertial systems. According to this standpoint, the replacement of the inertial systems by "Machian reference systems" is the content of the relativistic gravitational theory.

To support mathematical and physical arguments given in his 1913 paper with Grossmann (Ref. 27), Einstein [31] introduced the following argument in his "Comments". Consider a manifold with a region L containing none of the sources of the gravitational field (a "hole"), where accordingly the T_{ik} vanish. Then, by the matter outside of L, the metric g_{ik} is everywhere completely determined, also in L. Now, instead of the original coordinates x^i, we introduce new coordinates $x^{i'}$ of the following kind. Outside of L they be identical to x^i, but in L be $x^i \neq x^{i'}$. Of course, by such a substitution one can, at least for a part of L, reach that $g_{ik}' \neq g_{ik}$. Otherwise, it holds everywhere $T_{ik}' = T_{ik}$, namely outside of L for we have $x^{i'} = x^i$ in this region and in L because $T_{ik} = 0 = T_{ik}'$ holds. Hence, when arbitrary coordinate transformations are admitted, then in the case here under consideration more than one solution g_{ik} belongs to a given T_{ik}. That is Einstein's argument so far; it shows that Einstein had considered the solution g_{ik}' to be physically distinct from the solution g_{ik} and that he therefore believed generally covariant field equations not to provide unique solutions of a physical problem. Only after over two years he realised that the two solutions are mathematically distinct representations of the same physical situation.[9]

Einstein's "hole" argument does not only show the mathematical difficulties Einstein had to contend with at this time, but it makes also clear that he was unsatisfied by the fact that, in a theory with covariant gravitational equations, Mach's principle is not satisfied. In Einstein's reading of Mach, the totality of masses induces a g_{ik} field (the gravitational field) which itself governs all physical processes, including the propagation of light rays and the behaviour of length and time etalons, i.e., measurement rods and watches (Einstein [30]). That means that matter should also determine the reference systems definable by light, rods, and watches. However, in a theory of general relativity one can chose everywhere arbitrary reference coordinates and reference tetrads, respectively, both in L and outside of L: The matter described by T_{ik} does not fix either the reference systems or the gravitational field uniquely.

---

[8] For this, cf. Einstein's letters to Mach, Lorentz, Ehrenfest, and Besso in Ref. 26 (cf., *CP* 5, *No.* 470, 481, 484, 496, and 519).

[9] The point is that one should not forget that the terms "coordinate transformation" and, more generally, "transformations of the reference systems" here have quite a different meaning from that one in a space-time like the Minkowski world with specified space-time events which can be ascribed different coordinate values by certain transformations. One is working in a Riemannian space-time where, in the words of Stachel [66], one finds the following situation: "The physical properties of the points of the manifold inside the hole of any solution to the field equation depend *entirely* on the nature of the solution; hence, when one creates the second solution by dragging the physical significance is dragged along with the field. ... Until one introduces a metric, a manifold does not consist of physical events but just of mathematical points with *no* physical properties; it is the metric that imparts the physical character of events to the points of a manifold."

However, as long as one confines oneself to Riemannian spaces, such preferred reference systems can only exist when the gravitational equations themselves are not generally covariant, i.e., when Einstein's gravitational equations are replaced by non-coordinate-covariant or non-Lorentz-covariant equations, i.e., by equations that do not satisfy the principle of general relativity.[10] This early version of a relativistic gravitational theory can be considered as an attempt to formulate a theory lying between the two alternatives mentioned above. This theory, in some sense, realises a (local) field theoretical version of the Mach-Einstein doctrine saying that, via gravitation, the motion of matter described in concordance with the principle of equivalence determines the reference systems. (In this case, the reference systems are given by the coordinate systems.)[11]

Moreover, one can show [17] that there is a class of so-called Einstein-Grossmann theories stemming from the Lagrange density[12]

$$\begin{aligned} L_{EG} &= L_E + L^{*} \\ &= L_E + \sqrt{-g}[\alpha_1 \Gamma^{ikl}\Gamma_{ikl} + g^{ik}(\alpha_2 \Gamma^{l}{}_{il}\Gamma^{m}{}_{km} + \alpha_3 \Gamma^{l}{}_{ik}\Gamma^{m}{}_{lm} + \alpha_4 g_{mn} g^{rs}\Gamma^{m}{}_{rs}\Gamma^{n}{}_{ik})] \end{aligned} \tag{9}$$

where L_E denotes Einstein's Lagrange density (Einstein [34]) that, in the above sense, are Machian on the price of losing their general covariance. This class contains GRT (following from (9) for $\alpha_1 = \alpha_2 = \alpha_3 = \alpha_4 = 0$) as a degenerate case.

3. The fixing of the reference tetrads by the motion of cosmic matter

Except for all other possible arguments against the Einstein-Grossmann approach, it is, of course, unsatisfactory to work in a theory which is not generally covariant. Therefore, we follow another road to a theory satisfying a local version of the Mach-Einstein doctrine, namely the road to tetrad theories stemming from a Lagrangian built from Weitzenboeck's (anholonomic) invariants (Weitzenboeck [71]). That means, we shall work with a "Riemannian geometry with teleparallelism" leading us to a purely gravitational theory which is generally covariant and, nevertheless, satisfies the local version of the Mach-Einstein doctrine. In this case, the reference systems are given by the tetrads.

To avoid misunderstandings, it should be stressed here that Einstein himself in his papers on teleparallelism did not explicitly refer to Mach's principle. He (Einstein [35]) considered his "Riemannian geometry with teleparallelism" as an approach to a unified field theory of gravitation and electricity. In contrast to the gravitational theory "GRT", determining (up to gauge transformations) only the 10 algebraic combinations $g_{ik} = h^{A}{}_{i} h_{Ak}$ but not the 16 components of the tetrads $h^{A}{}_{i}$, the unification of gravitational and electromagnetic fields provides 16 Einstein-Maxwell-like equations which are able to deter-

[10] Of course, there is also the possibility to adhere to Einstein's equations and to supplement them by some conditions fixing the coordinates or reference systems, for instance, by requiring Fock's harmonic coordinates. But when these conditions are chosen once and for all as a fundamental of the theory then this amounts to a modification of Einstein's equations, too.

[11] For other formulations of Mach's principle, see the contributions in Ref. 6, and for the relation between the Mach-Einstein doctrine and the principles of equivalence and of general relativity see Ref. 70.

[12] In Riemannian space-time, it is the most general affine Lagrange density that is bilinear in the connection.

mine all 16 $h^A{}_i$ defining a teleparallelism. While in GRT the local reference systems are not connected with the reference systems in the neighbourhood now they are fixed and represent a lattice structure: The space-time structure of GRT representing a *Bezugs-Molluske* (reference mollusc) is replaced by a reference crystal permitting, instead of local, only global Lorentz rotations (Einstein [36], cf., also Ref. [68]). Hence, in Einstein's field theory with teleparallelism the additional equations fixing the reference systems are Maxwellian field equations but not Einstein-Mach conditions. Later, papers concerning such theories do however not refer to unified theories but discuss modifications of GRT mainly in order to win another theory of gravitation (cf., e.g., Ref. 59) for an early attempt).

It should also be mentioned that, motivated by the search for quantum gravitation, there are modern approaches to gravitation theory which go a step further and consider, instead of Riemannian spaces with teleparallelism, Riemann-Cartan spaces[13] where curvature and torsion are unequal to zero [43]. We however confine ourselves to teleparallelised Riemannian spaces[14] for the reason that we want to realise the Mach-Einstein doctrine within a theory satisfying the weak and the strong equivalence principle. The latter requirements, in particular, mean that the energy-momentum tensor of the matter sources is symmetric (see below) and that its covariant divergence vanishes. It is interesting to see that also this approach opens new perspectives for quantum gravitation, as will be demonstrated in the last part of this paper.

In Ref. 46 one finds a very general dynamical framework which is constructed in so-called metric-affine spaces admitting nonvanishing curvature, torsion, and nonmetricity. It embraces the Poincaré gauge field theories formulated in Riemann-Cartan spaces (with vanishing nonmetricity) as special case; and, in some sense, this is also true for theories working in teleparallelised Riemannian spaces like the theory with an Einstein-Mayer type Lagrangian here under consideration. In Ref. 54 it is argued that this theory is formally the same as the teleparallel special case of the Poincaré gauge field theory (to be precise, the author of this paper (H. Meyer) did not refer to Einstein-Mayer, but to Møller).

3.1 Field equations of Einstein-Mayer type

Let us start now, instead of from L_{EG}, from an Einstein-Mayer Lagrange density L_{EM} that is a scalar density with respect to coordinate transformations but not invariant

[13] Sometimes, such spaces are also called "Weyl-Cartan spaces".

[14] Cartan wrote in a letter to Einstein ([37], p.7): "In my terminology, spaces with a Euclidean connection allow of a *curvature* and a *torsion*: in the spaces where parallelism is defined in the Levi-Cività way, the torsion is zero, in the spaces where parallelism is absolute (*Fernparallelismus*) the curvature is zero, thus these are spaces without curvature and with torsion." Following this terminology, one could, if one thinks of their anholonomic representation, characterize Riemannian spaces with teleparallelism as spaces with nonvanishing torsion (given by the objects of anholonomy) and vanishing curvature (where the curvature tensor is constructed with the anholonomy objects). This view is natural when one starts with a Minkowski space having vanishing curvature and torsion. Then the relaxation of the Minkowskian condition of vanishing torsion leads to a Riemannian space with teleparallelism (in its anholonomic version) and the relaxation of both conditions leads to Riemann-Cartan spaces (see also Ref. 44).

under local Lorentz transformations. Similarly as in the above case, one finds here a restricted invariance, namely an invariance with respect to global Lorentz transformations. To discuss our question we shall confine ourselves to the Einstein-Mayer class of Lagrangians [35, 36, 38, 39, 40, 59]

$$L_{EM} = \sqrt{-g}R + ahF_{Aik}F^{Aik} + bh\phi_A\phi^A, \tag{10}$$

where $h = \det(h^A{}_i) = \sqrt{-g}$, $R = g^{ik}R_{ik}$ is the Ricci scalar, a and b are numerical constants, the ϕ_A are defined as $\phi_A = h^i{}_A\gamma^m{}_{im}$, and the F_{Aik} are Cartan's anholonomy objects

$$F_{Aik} = h_{Ai,k} - h_{Ak,i} = h_A{}^l(\gamma_{lik} - \gamma_{lki}) \tag{11}$$

(with the Ricci rotation coefficients $\gamma_{ikl} = h^A{}_i h_{Ak;l}$). The Lagrangian (10) lies on the basis of Møller's tetrad theory of gravitation (Møller [57]).[15] (From the viewpoint of a unified gravito-electromagnetic theory in the sense of Einstein's program of 1929, it is also discussed in Ref. 68; however, in contrast to Einstein's approach, there, like in Møller's theory an additional matter Lagrangian is introduced. For historical reasons, in the following we shall call the Lagrangian (10) "Einstein-Mayer Lagrangian", while the tetrad equivalent of the Einstein-Hilbert Lagrangian will be called "Møller Lagrangian". (For other metric-teleparallel theories, see, e.g., Kopczyński [52], Hayashi and Shirafuji [43]).

The Einstein-Mach theories based on (10) have the advantage that they are generally coordinate-covariant. Now, in the anholonomic (Einstein-Cartan) representation of the space-time structure the reference systems, i.e., the local orthonormal coframe $\vartheta^A = h^A{}_i(x^l)dx^i$ (the "tetrads" $h^A{}_i(x^l)dx^i$), are the gravitational field variables. From the standpoint of the theories here under consideration, Einstein's GRT is again a degenerate case, now following from (10) for $a = b = 0$.

In order to gain the possibility of interpreting the theory based on this Lagrangian as a macroscopic gravitational theory we have to supplement the Lagrangian (10) by a matter Lagrangian that couples matter minimally via the metric to gravitation. This introduces Hilbert's metrical energy-momentum tensor in the field equations

$$T_{ik} = \frac{2\delta}{\sqrt{-g}}\frac{L_{mat}}{\delta g^{ik}} \qquad \text{with } T_{ik} = T_{ki} \tag{12}$$

as source of the gravitational field [17, 18]. This way we arrive at field equations of the form

$$\frac{1}{h}\frac{\delta L_{EM}}{\delta h^A{}_i} = -\kappa h_A{}^l T^i{}_l \tag{13}$$

which also can be written as

$$\frac{1}{h}\frac{\delta L_{EM}}{\delta h^{Ai}} h^A{}_k = -\kappa T_{ik} \tag{14a}$$

with

$$\frac{\delta L_{EM}}{\delta h^{Ai}} h^A{}_k - \frac{\delta L_{EM}}{\delta h^{Ak}} h^A{}_i = 0. \tag{14b}$$

Due to the coordinate covariance of the Lagrangian, the Noether theorem leads with

$$\delta h_{Ai} = -h_{Ak}\delta x^k{}_{,i} - h_{Ai,k}\delta x^k \tag{15}$$

---

[15] In order to remove singularities from the theory of gravitation, later Møller [58] also introduced Lagrangians that are quadratic in the Weitzenboeck invariants.

to the identities [17]

$$\left(\frac{\delta L_{EM}}{\delta h_{Ai}} h_{Ak}\right)_{,i} - \frac{\delta L_{EM}}{\delta h_{Ai}} h_{Ai,k} = \left(\frac{\delta L_{EM}}{\delta h_{Ai}} h_{Ak}\right)_{;i} = 0. \qquad (16)$$

Before discussing this gravitational theory in more detail, let us compare it to the genuine Einstein-Mayer theory in order to make more evident what the introduction of a symmetric matter term means. [16] The variation of (10) with respect to the tetrads provides a tensor

$$G_{ik} = E_{ik} + \theta_{ik} \quad \text{with } E_{ik} = \text{Einstein tensor and } \theta_{ik} \neq \theta_{ki}. \qquad (17)$$

In the case of a unified geometric field theory, all physics had to result from a purely geometric Lagrangian. Then one had [36, 38, 53]

$$E_{ik} = -\theta_{(ik)} \quad \text{and} \quad \theta_{[ik]} = 0. \qquad (18)$$

Otherwise, by introducing a matter Lagrangian (Treder [68], Møller [57, 58]) one arrives by variation at the equations

$$G_{(ik)} = E_{ik} + \theta_{(ik)} = -\kappa T_{ik} \quad \text{and} \quad \theta_{[ik]} = 0 \quad (\text{with } T_{ik} \text{ given by (12)}). \qquad (19)$$

If one interprets them as 10+6 gravitational equations, then they determine, apart from arbitrary constant Lorentz rotations of the tetrads, the 16 gravitational potentials $h^A{}_i$. Together with the Noether identities (16), taking in this notation the form [58, 59]

$$H^i{}_{k,i} = F^i{}_{k;i} - F^{il}\gamma_{ilk} \quad \text{where} \quad H_{ik} \equiv E_{ik} + \theta_{(ik)} \text{ and } F_{ik} = \theta_{[ik]}, \qquad (20)$$

and the second equation in (19) this leads to the dynamical equations

$$T^k{}_{i;k} = 0. \qquad (21)$$

In the following let us confine ourselves to the case $b = 0$,[17] and accordingly assume the "Einstein-Larmor" Lagrangian supplemented by the matter Lagrangian L_{mat},

$$L = \sqrt{-g}(R + ahF_{Aik}F^{Aik} + 2\kappa L_{mat}). \qquad (22)$$

Then the Eqs. (14) can be written as four coupled Maxwell-like equations coupled by the currents $S_A{}^i$ [68]:

$$(\sqrt{-g}F_A{}^{ik})_{,k} = \frac{1}{2a} j_A{}^i \text{ with } j_A{}^i = \sqrt{-g}h_A{}^l(E^i{}_l + \kappa T^i{}_l + 2a[\tfrac{1}{4}\delta^i{}_l F_{Bmn}F^{Bmn} - F_{Blm}F^{Bim}]) \quad (23a)$$

or

$$F_A{}^{ik}{}_{;k} = \frac{1}{2a} S_A{}^i \quad \text{with} \quad S_A{}^i = h_A{}^l(E^i{}_l + \kappa T^i{}_l + +2a[\tfrac{1}{4}\delta^i{}_l F_{Bmn}F^{Bmn} - F_{Blm}F^{Bim}]) \quad (23b)$$

and, equivalently, the equations (23 a, b) take the form

$$E_{ik} = -\kappa T_{ik} - \theta_{ik} \qquad (24)$$

and

$$\theta_{ik} - \theta_{ki} = 0. \qquad (25)$$

Here $E_{ik} = R_{ik} - \frac{1}{2}g_{ik}R$ is the Einstein tensor, and the additional "matter tensor" θ_{ik} is given by[18]

$$\theta_{ik} = a(\tfrac{1}{2}g_{ik}F_{Amn}F^{Amn} - 2F_{Air}F^A{}_k{}^r + 2h^A{}_iF^A{}_k{}^l{}_{;l}). \qquad (26)$$

---

[16] For a discussion of this and of other mathematical and physical aspects of tetrad theories, see also Ref. 14.

[17] Einstein [35] and Levi-Cività [53] discussed this case as a candidate for a unified field theory.

[18] The Einstein tensor can be considered as defining, via the field equations, the total matter density. For instance, Schrödinger [64] wrote that he would not consider Einstein's GRT equations as field equations, but as the definition of the matter tensor.

Hence, the symmetry condition (25) implied by the above-mentioned assumption that the (symmetric) metrical energy-momentum tensor is presupposed to be the field source leads to six additional equations

$$\theta_{[ik]} = a(F_{Ak}{}^{l}{}_{;l}h^{A}{}_{i} - F_{Ai}{}^{l}{}_{;l}h^{A}{}_{k}) = 0 \tag{27}$$

fixing, as mentioned above, together with (24) the 16 tetrad components up to constant Lorentz transformations.

The four conservation laws satisfied by the "currents"

$$F_{A}{}^{ik}{}_{;ki} = \frac{1}{2a} S_{A}{}^{i}{}_{;i} = 0 \tag{28}$$

are equivalent to the Noether identities [68]. In virtue of the Bianchi identities $E_i{}^k{}_{;k} = 0$ and the condition (25), these identities imply that the additional matter term θ_{ik} is wattless [17]: $\theta_i{}^k{}_{;k} = 0$ such that the dynamical equations are satisfied (see also the letters between Einstein and Cartan 1929-1932 [37]).

To finish this subsection, we add the following remarks:

(1) The required symmetry of the matter tensor is an expression of the principle of equivalence: (i) According to the equivalence of the inertial and passive gravitational masses (tested by the Eötvös experiment), in a given gravitational field the special-relativistic conservation law of energy and momentum satisfied by the symmetric matter tensor $T_{ik}(=T_{ki})$ must be written generally covariantly with respect to the functions $g_{ik} = h^{A}{}_{i}h_{Ak}$ representing the inertial and gravitational potentials. This requires the dynamical equations (21) which now are really satisfied. (ii) According to the equivalence of the inertial and active gravitational masses (postulated in Newton's third law), the Poisson equation $\Delta\Phi = 4\pi G\rho$ (G is Newton's gravitational constant and ρ the active mass density) can and must be replaced by equations of the kind $W_{ik} = -\kappa T_{ik}$. These equations determine the g_{ik}, where, because of the equivalence of inertial and active gravitational masses, the symmetric matter tensor T_{ik} describing the inertial properties of matter[19] can replace the Newtonian ρ.

(2) Of course, Weyl [72, 73] was right when he criticised Einstein's unified field theory based on teleparallelism with the argument that it violates the ordinary conservation laws. However, this criticism is not justified in the case where one interprets the Einstein-Mayer equations as purely gravitational field equations satisfying the strong principle of equivalence.

(3) The theory here under consideration solves the Einstein "hole problem". Indeed, the motion of matter described in concordance with the principle of equivalence determines the reference systems so that only constant Lorentz rotations are admitted. Vice versa, this means that, if one performed a local Lorentz transformation leading to an inadmissible reference system one would produce another type of matter not satisfying the equivalence principle. For, since the field equations are not covariant under these transformations equation, (24) would define a new matter tensor (see Footnote 19) and, due to the invalidity of (25) in the new reference system, one had instead of (21) the relation

$$(T^{i}{}_{k} + \theta^{i}{}_{k})_{;i} = 0 \tag{29}$$

[19] This is true because the introduction of T_{ik} is based on the equivalence of energy and inertia.

saying that only the covariant divergence of the total matter tensor vanishes.

3.2 On the canonical formalism

A canonical formalism that is physically interpretable requires a Lagrangian which is a bilinear function of the first derivatives of the variables $h^A{}_i$ of the configuration space.[20] If one assumes the field quantities to be these variables, then, in the Einstein-Mayer Lagrangian, one has to replace the scalar density $\sqrt{-g}R$ by Møller's Lagrange density (Møller [56])

$$L_M = h(\gamma^{ilm}\gamma_{mli} - \phi_A\phi^A) \tag{30}$$

which is equal to $\sqrt{-g}R$ modulo the Euler variation. Then the Einstein-Larmor Lagrangian takes the form

$$L^* = h(\gamma^{ilm}\gamma_{mli} - \phi^A\phi_A) + ahF^{Aik}F_{Aik}. \tag{31}$$

As was stated, e.g., by Sławianowski [65], Lagrange densities of this type, or more general, Lagrange densities for the field of real tetrads, invariant under Diff$M \times G$, G being a certain, as yet unspecified, Lie subgroup of $GL(\mathrm{n}, \mathbf{R})$, lead to field equations which are formally similar to Maxwell equations[21, 22]

$$H_A{}^{ik}{}_{,k} = j_A \qquad \text{with } H_A{}^{ik} = -H_A{}^{ki}\frac{\partial L}{\partial h^A{}_{i,k}} \quad \text{and} \quad j^i{}_A = \frac{\partial L}{\partial h^A{}_i}. \tag{32}$$

Here the source terms do not correspond to external fields, but are self-interaction currents resulting from the nonlinear violation of the gradient invariance. The equations (23a, b) discussed above are a special version of (32) resulting if one assumes the Lagrangian (10) or (31) which are invariant under the global Lorentz group $O(1, 3)$. Therefore, most of the following statements hold also in the more general case (32) discussed in [65]. [In parentheses, it should be mentioned that in our special case the field equations (23b) show that, although the source term is a self-interaction current, one can interpret these equations as determining the anholonomy objects with respect to a Riemannian metric.]

The inspection of the field equations (23a, b) shows that four of them, namely[23]

$$F_A{}^{0k}{}_{,k} = \frac{1}{2a}j_A{}^0 \quad \text{or} \quad F_A{}^{0k}{}_{;k} = \frac{1}{2a}S_A{}^0, \tag{33}$$

are Lagrangian constraints:

$$\varphi_A := F_A{}^{0k}{}_{;k} - \frac{1}{2a}S_A{}^0 = 0. \tag{34}$$

There are no further constraints since, due to the field equations and the constraints themselves, one obtains $\partial_0\varphi_A = 0$. At least for pure gravitation, one finds here a similar situation as in the case of the Einstein-Maxwell theory: The field equations are in involu-

---

[20] Otherwise, the *00*-component of the energy-momentum tensor cannot be identified with the Hamiltonian.

[21] Our notation differs from that one in Ref. 65.

[22] As was shown by Hehl in Ref. 44 (and later papers [45-47]), also the metric-affine theories take such a form, or more precisely, a "quasi-Yang-Mills form".

[23] We assume here manifold M which is topologically $M = \Sigma \times R$ and a choice of the coordinates where the time vector of the (3+1)-foliation is tangential to the time-coordinate curves.

tion. However, it there is of course a great difference to the Einstein-Maxwell theory for the gravitational and the Maxwell field are described by independent variables.

As to the Hamiltonian constraints which one needs in order to follow the Bergmann-Dirac algorithm of canonical quantisation (Bergmann [1, 8-11], Dirac [22-24])[24], first this implies the impossibility to use the formulation proposed by Ashtekar [2-4] in which a real densitised triad (tangential to the spacelike hypersurfaces introduced by the foliation) and a connection 1-form [which takes values in the complexified Lie algebra of $SO(3)$] are the canonical variables. Indeed, any attempt of a (3+1)-formulation of the Lagrangian leads, for the Møller part of the Lagrange density, to a time-directed Lie derivative of the above-mentioned Ashtekar connection and, for the Maxwell part, to a Lie derivative of the triads. This is of course due to the fact that, in the Maxwell term, the tetrads and triads, respectively, take the part of the connection 1-form (with values in the Lie algebra of $U(1)\times U(1)\times U(1)\times U(1)$).[25]

Assuming, for this reason, the tetrads $h^A{}_i$ to be the variables of the configuration space, then four of the momenta conjugate to these variables vanish so that one is led to the primary constraints:

$$\pi^0{}_A = \frac{\partial L^*}{\partial h^0{}_{A,0}} = 0. \tag{35}$$

(Furthermore, there will occur also secondary constraints). To construct the Hamiltonian one can follow the procedure introduced by Goldberg [42] for the Lagrangian (30). It uses triad densities as basic configuration-space variables where, with the lapse and the shift functions, the triad can be enlarged to an orthogonal tetrad whose pullback is the triad. In the anholonomic version of GRT resulting from (30) one gets 3+1+3 constraints: three constraints corresponding to changes of the canonical variables resulting from spatial diffeomorphisms on Σ, one constraint corresponding to the time-evolution of the initial data, and three constraints corresponding to changes stemming from the $SO(3)$ gauge transformations (the latter ones appear additionally to the four first ones which are due to the coordinate-covariance and therefore already known from the holonomic representation of GRT). In the case (22), where the Lagrangian is not Lorentz-covariant, the major difference to the anholonomic version of GRT will be that there occur only the four constraints corresponding to the coordinate-covariance, while the latter constraints are replaced by the field equations (27).

4. Conclusions: On new perspectives in quantum gravitation

Under the assumption that the strong principle of equivalence holds, the theoretical realisation of the Mach principle (in the local version of the Mach-Einstein doctrine) and of the principle of general relativity are alternative programs. That means, only the for-

---

[24] See also the review article by Wipf [74].

[25] One finds of course quite another situation if one couples a Maxwell or Yang-Mills field to the gravitational field, where the gravitational and the matter field are described by independent variables (cf., e.g., Ref. [61]).

mer or the latter can be realised – at least, as long as the Riemannian framework is not given up and only field equations of second order are considered.

In the anholonomic representation of the space-time referred to in the previous section, one finds Møller's Lagrangian (30) which, by multiplication with $h^A{}_i$, can also be written as,

$$L_M = h(\gamma_{ABC}\gamma^{CBA} - \gamma_A{}^{BA}\gamma^C{}_{BC}) \tag{36}$$

which is a coordinate-covariant scalar density but not covariant with respect to local Lorentz transformations of the reference systems given by the tetrads. Therefore, in GRT the energy-momentum density of the gravitational field is not localisable. Indeed, Einstein's energy-momentum complex (Einstein [34])

$$\sqrt{-g}(t^k{}_i)_E = \frac{1}{2\kappa}\left(\frac{\partial L_E}{\partial g^{mn}{}_{,k}} g^{mn}{}_{,i} - \delta^k{}_i L_E\right) \tag{37}$$

corresponding to $\sqrt{-g}R$ is a Lorentz invariant expression but no coordinate-covariant tensor density. Otherwise, Møller's complex

$$\sqrt{-g}(t^k{}_i)_M = \frac{1}{2\kappa}\left(\frac{\partial L_M}{\partial h^A{}_{m,k}} h^A{}_{m,i} - \delta^k{}_i L_M\right) \tag{38}$$

attributed to the Lagrangian (30) is a coordinate-covariant tensor density but not covariant with respect to local Lorentz transformations of the tetrads. Therefore, one can give the Einstein complex an arbitrary value by an appropriate choice of the space-time coordinate system, and, similarly, the Møller complex can be given an arbitrary value by an appropriate choice of the reference systems, i.e., of the anholonomic coordinates. As to the Einstein complex, this was already mentioned by Bauer [7], Schrödinger [63], and Freud [41], and for the Møller complex it was mentioned by Møller himself (Møller [56]). As a consequence, Møller proposed the equations mentioned above.

But, if one cannot define a localisable energy-momentum of the gravitational field, then a quantisation of this field according to the rules of the canonical quantisation meets great problems (see Refs. 5, 50, 51). In particular, one cannot identify the *00*-component of the energy-momentum complex as the Hamiltonian. As was argued in Refs. 13, 15 16, 20, these problems signal that, according to GRT, there are no measurable quantum effects of the gravitational field or, in other words, the general principle of relativity and quantum gravitation are not compatible.

However, in a Riemannian field theory with Einstein-Cartan teleparallelism resting on the Lagrangian (31) one finds quite a different situation. Here one has a Lagrangian of canonical structure that is coordinate-covariant but, like Møller's Lagrangian, not covariant with respect to local Lorentz transformations. But now the missing Lorentz covariance is intended. The Lagrangian is to lead to field equations not satisfying the general principle of relativity but to equations fixing the 16 components of the tetrads $h^A{}_i$ instead of their 10 combinations $g_{ik} = h^A{}_i h_{Ak}$, and the Hamiltonian $\boldsymbol{H}$ is given by the *00*-component of the energy-momentum complex (which is a tensor with respect to the group of global Lorentz transformations lying on the basis of this theory),

$$\begin{aligned} \boldsymbol{H} &= \frac{1}{\kappa}\left(\frac{\partial \boldsymbol{L}^{\bullet}}{\partial h^{Aa}{}_{,0}} h^{Aa}{}_{,0} - \boldsymbol{L}^{\bullet}\right) \\ &= \frac{h}{\kappa}\left[(h^{0}{}_{A;a} - \delta^{0}{}_{a} h^{m}{}_{A;m}) h^{Aa}{}_{,0} - (\gamma^{ilm}\gamma_{mli} - \phi^{A}\phi_{A})\right] + 2a(F_{A}{}^{0}{}_{r} h^{Ar}{}_{,0} - \frac{1}{2} F^{Amn} F_{Amn}) \quad (39) \\ &= 2(t^{0}{}_{0})_{M} + \frac{2ah}{\kappa}(F_{A}{}^{0}{}_{r} h^{Ar}{}_{,0} - \frac{1}{2} F^{Amn} F_{Amn}) \end{aligned}$$

In (39) the term $2(t^{0}{}_{0})_{M}$ is the Hamiltonian of the Riemannian geometrical field g_{ik} and

$$\boldsymbol{H}_{F} = \frac{2ah}{\kappa}(F_{A}{}^{0}{}_{r} h^{Ar}{}_{,0} - \frac{1}{2} F^{Amn} F_{Amn}) \tag{40}$$

is the Hamiltonian of the (Einstein-Cartan) torsion field F_{Aik}. According to this Einstein-Mayer theory of gravitation, the energy-momentum density of gravitational fields is a localised and measurable tensor density.[26] As a consequence of the breaking of local Lorentz covariance, in this theory measurable quantum effects of gravitation should exist, namely as "Machian effects" with reference to the universe. Thus, gravitons are cosmically induced Goldstone particles (Heisenberg [48], cf. also Ref. 67).

In quantum electrodynamics, the quantisation of the Maxwell field leads to Heisenberg-Pauli uncertainty relations limiting the simultaneous measurement of the electric field strength $\boldsymbol{E}$ and the vector potential $\boldsymbol{A}$

$$\Delta\boldsymbol{E}\Delta\boldsymbol{A} \geq \hbar c \quad \text{or} \quad \Delta\boldsymbol{E}\Delta\boldsymbol{A} \geq \alpha^{-1} e^{2} \tag{41}$$

where $\hbar$ denotes Planck's quantum action and $\alpha = \frac{e^2}{\hbar c}$ Sommerfeld's constant. Analogously, the quantisation of the Maxwell-like equations (23a, b) implies the uncertainty relations

$$\frac{G}{c^4}\Delta F_{A}\Delta h^{A}(\Delta x)^{3} \geq \hbar c \quad \text{or} \quad \Delta F_{A}\Delta h^{A}(\Delta x)^{3} \geq \frac{\hbar G}{c^{3}} \tag{42}$$

(where $\sqrt{\hbar G/c^{3}}$ is Planck's length) corresponding to the commutation rules

$$[F_{A}, h^{A}] = \delta^{3}(x)\frac{\hbar G}{c^{3}}. \tag{43}$$

The existence of these relations does not mean a quantisation of the Riemannian metrical structures of the space time but rather the quantisation of the teleparallelism introduced in the space-time: Not the metrical tensor (as for instance, in the ADM formalism, its projection on spacelike hypersurfaces) and certain concomitants of the Christoffel connection (like the external curvature of those surfaces) are the canonically conjugate variables but rather the frame vectors $h^{A}{}_{i}$ and the anholonomy objects (Einstein-Cartan) torsion field $F^{A}{}_{ik}$.

For the purely metric part one has again the Rosenfeld inaccuracy relations (Rosenfeld [62]) saying that there are no measurable quantum-gravitational effects that can be ascribed to this part. But these relations do not correspond to commutation rules since they relate gravitational field quantities to space-time coordinates or intervals.

Rosenfeld's relations can be derived from the principle of equivalence, Heisenberg's "fourth uncertainty relation", and Bohr's nonlocality condition. Indeed, by taking the average of Einstein's equations

$$E_{ik} = -\kappa \tilde{T}_{ik} \tag{44}$$

---

[26] It remains however the task to test whether the energy defined in this way can be shown to be positive-definite or whether, due to the general covariance, one meets again the difficulties known from GRT.

over the volume $L_0^3 \equiv (\Delta x)^3$ of the measurement region (given by the extent of the measurement body) one gets for the uncertainties (or inaccuracies) ΔR and ΔT of the quantities R_{ik} and $\tilde{T}_{ik}$ the relation[27]

$$\Delta R \approx \kappa \Delta \tilde{T} \approx \kappa \Delta E (\Delta x)^{-3} \tag{45}$$

and thus for the uncertainties Δg of g_{ik},

$$\Delta g \approx \Delta R (\Delta x)^2 \approx \kappa \Delta E (\Delta x)^{-1}, \tag{46}$$

where ΔE is the uncertainty of E. Hence, Heisenberg's fourth relation

$$\Delta E \Delta t \geq \hbar \tag{47}$$

and Bohr's nonlocality condition to be imposed on an optimal measurement body

$$\Delta x \geq c \Delta t \tag{48}$$

yield

$$\Delta g (\Delta x)^2 \approx \kappa \Delta E \Delta x \geq \kappa c \Delta E \Delta t \geq \kappa \hbar c \tag{49}$$

and thus the Rosenfeld inequality

$$\Delta g_{ik} \Delta x^i \Delta x^k \geq \kappa \hbar c, \tag{50}$$

where $(\kappa \hbar c)^{1/2}$ is Planck's fundamental length.

Relation (50) must be compatible with each quantisation procedure applied to such fields. If one assumes Einstein's condition that the Lorentz signature of space-time has to be maintained, i. e., that $\Delta g \leq 1$, then Rosenfeld's relation (50) says that $(\Delta x) \geq \kappa \hbar c$, i. e., that Planck's length is the smallest measurable distance. According to Rosenfeld, relation (50) (possibly) shows that, at distances of the order of magnitude of Planck's length, the quantum fluctuations induced by the measuring procedure are so violent that the quantum gravity region is not accessible to measurement. This interpretation is especially plausible when one follows Rosenfeld's own "measurement-theoretical" derivation of (50).

They are due to the occurrence of planckions as smallest test bodies and biggest elementary particles. They become evident in a "topological quantisation" of the anholonomic representation of the Riemann space-time by means of the integral quantisation rules of Bohr and Sommerfeld (discussed, e.g., in Ref. 20).

5. Acknowledgment. We cordially thank Jan J. Sławianowski for useful suggestions for improvements of this paper.

6. References

1. J. L. Anderson and P. G. Bergmann, *Phys Rev.* **83** (1951) 1018.
2. A. Ashtekar, *Phys. Rev.* **D36** (1987) 1587.
3. A. Ashtekar, *New Perspectives in Canonical Gravity* (with invited contributions) (Bibliopolis, Naples, 1988).
4. A. Ashtekar, *Lectures on Non-Perturbative Canonical Gravity* (Notes prepared in

[27] Here the ≈ signs relate the usual (stochastic) measurement errors, while ≥ denotes inequalities arising from fundamental relations of general relativity, quantum field, and measurement theories.

collaboration with R. S. Tate) (World Scientific, Singapore, 1991).
5. A. Ashtekar and J. Stachel (eds.), *Conceptual Problems of Quantum Gravity* (Birkhäuser, Boston, 1991).
6. J. Barbour and H. Pfister, (eds.), *Mach's Principle: From Newton's Bucket to Quantum Gravity* (Birkhäuser, Boston etc., 1995).
7. H. Bauer, *Phys. Z.* **19**, (1918) 163.
8. P. G. Bergmann, *Phys. Rev.* **75** (1949) 680.
9. P. G. Bergmann, *Rev. Mod. Phys.* **33** (1961) 510.
10. P. G. Bergmann, "Status of Canonical Quantization", in *Relativity and Gravitation*, ed. C. Kuper and A. Peres (Gordon and Breach, New York, 1971).
11. P. G. Bergmann and A. Komar, *Phys. Rev. Lett.* **4** (1960) 432.
12. N. Bohr and L. Rosenfeld, "Zur Frage der Meßbarkeit der elektromagnetischen Feldgrößen", *Dan. Vid. Selskab. Mat-fys.* **12** (1933) No. 8. [English translation: "On the Question of the Measurability of Electromagnetic Field Quantities", in *Niels Bohr, Collected Works*, Vol. 7, ed. J. Kalckar (North-Holland, Amsterdam et al., 1996)].
13. H.-H. v. Borzeszkowski, "On Self-Interaction and (Non-)Perturbative Quantum GRT", in *The Present Status of the Quantum Theory of Light*, ed. S. Jeffers et al. (Kluwer Academic Publishers, Dordrecht, 1997).
14. H.-H. v. Borzeszkowski, U. Kasper, E. Kreisel, D.-E. Liebscher, and H.-J. Treder, *Gravitationstheorie und Äquivalenzprinzip*, ed. H.-J. Treder (Akademie-Verlag, Berlin, 1971).
15. H.-H. v. Borzeszkowski and H.-J, Treder, *The Meaning of Quantum Gravity* (Reidel, Dordrecht, 1988).
16. H.-H. v. Borzeszkowski and H.-J, Treder, *Found. Phys.* **25** (1993) 291.
17. H.-H. v. Borzeszkowski and H.-J, Treder, *Found. Phys.* **26** (1996) 929.
18. H.-H. v. Borzeszkowski and H.-J, Treder, *Found. Phys.* **27** (1997) 595.
19. H.-H. v. Borzeszkowski and H.-J, Treder, *Found. Phys.* **28** (1998) 273.
20. H.-H. v. Borzeszkowski, V. de Sabbata, C. Sivaram, and H.-J. Treder, *Found. Phys. Lett.* **9** (1996) 157.
21. P. A. M. Dirac, *Canad. J. Math.* **2** (1950) 129.
22. P. A. M. Dirac, *Canad. J. Math.* **3** (1951) 1.
23. P. A. M. Dirac, *Proc. R. Soc. (Lond.)* **A346** (1958) 326.
24. P. A. M. Dirac, *Lectures on Quantum Mechanics* (Yeshiva University, New York, 1964).
25. A. Einstein, *The Collected Papers of Albert Einstein*, Vol. 4, "The Swiss Years: Writings 1912 - 1914", ed. M. J. Klein, A. J. Kox, J. Renn, and R. Schulmann (Princeton University Press, Princeton, 1995).
26. A. Einstein, *The Collected Papers of Albert Einstein*, Vol. 5, "Correspondence 1902 - 1914", ed. M. J. Klein, A. J. Kox, and R. Schulmann (Princeton University Press, Princeton, 1993).
27. A. Einstein and M. Grossmann, *Entwurf einer verallgemeinerten Relativitätstheorie und einer Theorie der Gravitation* (Teubner, Leipzig, 1913); cf. also *CP* **4**, No.

13.[28]
28. A. Einstein, *Phys. Z.* **XIV** (1913) 1249; cf. also *CP* **4**, *No.* 17.
29. A. Einstein, *Phys. Z.* **XV** (1913) 174; cf. also *CP* **4**, *No.* 25.
30. A. Einstein, Letters to E. Mach, H. A. Lorentz, P. Ehrenfest, and M. Besso, in *Collected Papers* **5**, No. 495, No. 467, No. 481, 484, and No. 516.
31. A. Einstein, "Bemerkungen" (Appendix to the reprint of the Einstein-Grossmann paper), *Z. für Math. u. Phys.* **62** (1914) 260; cf. also *CP* **4**, No. 26.
32. A. Einstein, *Scientia* **15** (1914) 328; cf., also *CP* **4**, *No. 31.*
33. A. Einstein, "Formale Grundlage der allgemeinen Relativitätstheorie", *Berliner Ber.* 1914, p. 1030.
34. A. Einstein, "Der Energiesatz in der allgemeinen Relativitätstheorie", *Berliner Ber.* 1918, p. 448.
35. A. Einstein, "Riemann-Geometrie unter Aufrechterhaltung des Begriffs des Fernparallelismus", *Berliner Ber.* 1928, p. 219.
36. A. Einstein, "Einheitliche Feldtheorie und Hamiltonsches Prinzip", *Berliner Ber.* 1929, p. 124.
37. A. Einstein and E. Cartan, in *Elie Cartan - Albert Einstein, Letters on Absolute Parallelism, 1929 - 1932*, ed. R. Debever (Princeton University Press, Princeton and New Jersey, 1979).
38. A. Einstein and W. Mayer, "Systematische Untersuchung über kompatible Feldgleichungen, welche von einem Riemannschen Raum mit Fernparallelismus gesetzt werden können", *Berliner Ber.* 1931, p. 3.
39. A. Einstein, "Neue Möglichkeit für eine einheitliche Feldtheorie von Gravitation und Elektrizität", *Berliner Ber.* 1928, p. 224.
40. A. Einstein, "Zur einheitlichen Feldtheorie", *Berliner Ber.* 1929, p. 2.
41. Ph. Freud, "Über die Ausdrücke der Gesamtenergie und des Gesamtimpulses eines materiellen Systems in der allgemeinen Relativitätstheorie", *Ann. Math. Princeton* **40** (1939) 417.
42. J. N. Goldberg, *Phys. Rev.* **D 37** (1988) 2116.
43. K. Hayashi and T. Shirafuji, *Phys. Rev.* **D 19** (1979) 3524.
44. F. W. Hehl, *Found. Phys.* **15** (1985) 451.
45. F. W. Hehl and J. D. McCrea, *Found. Phys.* **16** (1986) 267.
46. F. W. Hehl, J. D. McCrea, E. W. Mielke, and Y. Ne'eman, "Metric-Affine Gauge Theory of Gravity: Field Equations, Noether Identities, World Spinors, and Breaking of Dilation Invariance", *Physics Reports* **258** (1995) pp. 1 - 171.
47. F. W. Hehl and F. Gronwald, "On Gauge Aspects of Gravity", in *Quantum Gravity,* ed. P. G. Bergmann, V. de Sabbata, and H.-J. Treder (World Scientific, Singapore, 1996).
48. W. Heisenberg, *Einführung in die einheitliche Feldtheorie der Elementarteilchen* (Hirzel, Stuttgart, 1967).
49. H. Hertz, *Die Prinzipien der Mechanik* (Barth, Leipzig, 1894). [The English translation was published by Macmillan in 1900.]

---

[28] *CP* stands for *Collected Papers of Albert Einstein.*

50. C. J. Isham, "Prima Facie Questions in Quantum Gravity", in *Canonical Gravity: From Canonical to Quantum*, ed. J. Ehlers and H. Friedrich (Springer, Berlin and Heidelberg, 1994), pp. 1-21.
51. C. J. Isham, "Conceptual and geometrical problems in quantum gravity", in *Recent Aspects of Quantum Fields*, ed. H. Mitter and H. Gausterer (Springer, Berlin, 1991).
52. W. Kopczyński, *J. Phys.* A **15** (1982) 493.
53. T. Levi-Cività, "Vereinfachte Herstellung der Einsteinschen einheitlichen Feldgleichungen", *Berliner Ber.* 1929, p. 2.
54. H. Meyer, *Gen. Rel. Grav.* **14** (1982) 531.
55. C. Møller, *Math.-Fys. Skr. Dan. Vid. Selskab* **1** (1961) No. 10.
56. C. Møller, "Survey of Investigations on the Energy-Momentum Complex in General Relativity", in *Entstehung, Entwicklung und Perspektiven der Einsteinschen Gravitationstheorie*, ed. H.-J. Treder (Akademie-Verlag, Berlin, 1966).
57. C. Møller, *Math.-Fys. Skr. Dan. Vid. Selskab* **39** (1978) No. 13.
58. C. Møller, "Are the Singularities in the Theory of Gravitation Inevitable?", in *Einstein-Centenarium 1979*, ed. H.-J. Treder (Akademie-Verlag, Berlin, 1979).
59. C. Pellegrini and J. Plebański, "Tetrad Fields and Gravitational Fields", *Math.-Fys. Skr. Dan. Vid. Selskab* **2** (1963) No. 4.
60. M. Planck, *Das Prinzip von der Erhaltung der Energie*, 2nd ed. (Teubner, Leipzig, 1913).
61. J. D. Romano, *Gen. Rel. Grav.* **25** (1993) 759.
62. L. Rosenfeld, "Quanten und Gravitation", in *Entstehung, Entwicklung und Perspektiven der Einsteinschen Gravitationstheorie*, ed. H.-J. Treder (Akademie-Verlag, Berlin, 1966).
63. E. Schrödinger, "Energiekomponenten des Gravitationfeldes", *Phys. Z.* **19** (1918) 430.
64. E. Schrödinger, *Space-Time Structure* (Cambridge University Press, Cambridge, 1950).
65. J. J. Sławianowski, *Nuovo Cimento* **106 B** (1991) 645.
66. J. Stachel, "Einstein and Quantum Mechanics", in *Conceptual Problems of Quantum Gravity*, ed. A. Ashtekar and J. Stachel (Birkhäuser, Boston etc., 1991).
67. H.-J. Treder, *Found. Phys.* **1** (1970) 37.
68. H.-J. Treder, *Ann. Phys. Leipzig* **35** (1978) 371.
69. H.-J. Treder, "On the Connection between Pico- and Mega-Cosmos", in *Old and New Questions in Physics, Cosmology, Philosophy, and Theoretical Biology. Essays in Honor of Wolfgang Yourgrau*, ed. A. van der Merwe (Plenum, New York, 1983) pp. 37-51.
70. H.-J. Treder, H.-H. von Borzeszkowski, A. van der Merwe, and W. Yourgrau, *Fundamental Principles of General Relativity Theories. Local and Global Aspects of Gravitation and Cosmology* (Plenum Press, New York and London, 1980).
71. R. Weitzenboeck, "Differentialinvarianten in der Einsteinschen Theorie des Fernparallelismus", *Berliner Ber.* 1928, p. 466.
72. H. Weyl, *Zs. Physik* **56** (1929) 330; cf. also Hermann Weyl, *Gesammelte*

Abhandlungen (Springer, Berlin, 1968) Vol. III, No 85.

73. H. Weyl, *Naturwiss.* **19** (1931) 49; cf. also Hermann Weyl, *Gesammelte Abhandlungen* (Springer, Berlin, 1968) Vol. III, No 93.
74. A. Wipf, "Hamilton's Formalism for Systems with Constraints", in *Canonical Gravity: From Canonical to Quantum*, ed. J. Ehlers and H. Friedrich (Springer, Berlin, Heidelberg, 1994) pp. 22-58.

NON-LOCALITY VERSUS LOCALITY: THE EPISTEMOLOGICAL BACKGROUND OF THE EINSTEIN-BOHR DEBATE

HORST-HEINO v. BORZESZKOWSKI
Technical University Berlin, Institute for Theoretical Physics, Hardenbergstr. 36, D-10623 Berlin; e-mail: borzeszk@itp.physik.tu-berlin.de

and

RENATE WAHSNER
Max Planck Institute for the History of Science, Wilhelmstr. 44, D-10117 Berlin; e-mail: wahsner@mpiwg-berlin.mpg.de

ABSTRACT

In its core, the non-locality–locality debate is a discussion about the concept of the physical reality. The difference between the measurement bases of classical and quantum mechanics is often interpreted as a loss of reality arising in quantum mechanics. In this paper it is shown that this apparent loss occurs only if one believes that everyday experience determines the Euclidean space, instead of considering this space, both in classical and quantum mechanics, as a theoretical construction needed for measurement and representing one part of a dualistic space conception. From this point of view, Einstein's program of a unified field theory can be interpreted as the attempt to find a physical theory that is less dualistic. However, if one regards this dualism as resulting from the requirements of measurements, one can hope for a weakening of the dualism but not expect to remove it completely. Further, it follows from the requirements of measurements that the Heisenberg cut cannot be arbitrarily far moved towards the observer.

Epistemological discussions on quantum mechanics yield two insights: (i) To acquire physical knowledge, means of cognition are always necessary, and measurement provides this means. (ii) The space of measurement and the space of the representation of physical dynamics must be distinguished conceptually; both coincide factually in classical physics, but not so in quantum mechanics.

1. Introduction

Discussions on the epistemological problems raised by quantum mechanics derive explicitly or implicitly from papers published by the founders of quantum theory, and the difficulties with which they are concerned are already known from those classic papers. Many of these problems, including those ones on the non-locality problem, address the question whether or not a conception of reality can be established that will play the same role entertained by classical mechanics but without contradicting quantum mechanics. On the whole, such discussions start with the assumption that one (or, better, the physicists of the last century) could interpret classical mechanics as a description of "what really is". In doing so, one disregards an argument going back to antiquity: that we know nothing about what anything really is in itself, that – as Democritus claimed – the ultimate truth remains deeply in the dark. Modern-Age version of this thought was formulated by Kant.[1] And indeed, discussing the problem of physical reality in today, one must not forget Kant's epistemological turn. However, the fact that Kant was led to his philosophical system by analysing Newtonian mechanics is often ignored in this discussion. And if it is not ignored, then it is only seen as a philosophical doctrine partly derived from physics but external to physics. It is not viewed as necessary to clarify the status of classical mechanics and, thus, of its concept of reality.

To our mind, it is just this starting point that leads in quantum physics to conceptions of physical reality that run into difficulties. Our point is that even classical mechanics could never have been interpreted as describing "reality in itself". It is also a *theory* (i.e., not identical to reality) that is related to nature by measurement – philosophically speaking, by means of cognition (*Erkenntnismittel*). And we think that the essential problems one encounters in understanding of the status of quantum mechanics result from failing to keep in mind this epistemological basis of classical mechanics (and of physics in general).

Indeed, if one believes that it is possible to consider classical mechanics as a description of reality in itself, then the fundamental role of measurement in quantum mechanics will appear as a disaster, undermining (or at least weakening) the objectivity of physics, instead of as a instance of the general rule: that each physical theory requires means of cognition that must be reflected together with the theory under consideration. This results in a split between classical and quantum mechanics – a barrier to the understanding of quantum mechanics, which arises from a misunderstanding of classical mechanics.

The purpose of our paper is to analyse the problem of reality in quantum mechanics from the point of view that the epistemological reflection of physics must consider the means of cognition (in physics: the means of measurement) mediating between the subject and the object of cognition. In doing so, one crosses, of course, the border between physics and philosophy. There is, however, no objection to such a consideration because one must be ready to do that: One cannot seek to answer genuinely philosophical questions in physics if one simultaneously aims to exclude philosophy from consideration. In this analysis we are guided by the famous discussions between Bohr, Einstein, Heisenberg, and Schrödinger.

To solve problems concerning the quantum-mechanical conception of reality, one has also, as suggested above, to reconsider concepts of measurement and reality in classical mechanics. But there is a second reason to refer to classical mechanics. For the latter is not only the most important theory to which quantum mechanics should be compared, but it is simultaneously an essential part of the measurement basis for the whole edifice of physics (and thus also for quantum mechanics). It is just the latter aspect that was

---

[1] See Schrödinger in Ref. 36, p. 163.

emphasised by some authors when, referring to the meaning of classical mechanics for measurement, they called it "protophysics".[2]

2. Quantum Mechanics and the Physical Reality Concept

Many scientists feel epistemological problems brought up by quantum mechanics to be a "scandal of cognition". Niels Bohr expressed these feelings in the words: "For those who are not shocked when they first come across quantum theory cannot possibly have understood it."[3] He himself was shocked – even though only for a time. His great antagonist in the dispute about the new physics, Albert Einstein, was so until the end of his life. Of course, as founders of quantum mechanics both accepted this theory, but they appraised it differently with respect to its place in the physical process of cognition. The discussion of the so-called Einstein-Podolsky-Rosen paradox makes this exemplary clear [15]. Einstein admits that, from the standpoint of quantum mechanics alone, a paradoxical situation does not occur. Bohr says the same, but he means: Therefore everything is all right. Einstein answers: Not exactly that; the quantum-mechanical conception cannot be physic's last word. "In deep down it is wrong even so it is correct empirically and logically."[4]

One could be shocked by quantum mechanics because of its great difference to classical mechanics. However, that this difference was so striking is not due so much to the factual difference between these two sciences, but to an inadequate awareness of the epistemological foundation of classical mechanics and physics in general. This inadequate consciousness became clearly indefensible in the case of quantum mechanics. For quantum mechanics reveals that physics needs a means of cognition in the form of theoretically thought and technically realised measurement means which, in its function, should be thought jointly with the object dealing with. One must do so if one is to conceive the content of the physical theory under consideration. Understanding quantum mechanics (or the difference between classical and quantum mechanics) means grasping this insight as essential to the understanding of each branch of physics, including that of classical mechanics, and not considering it only as a specific feature of quantum mechanics. Bohr's and Einstein's famous dialogue on quantum mechanics shows that the "scandal" of physical cognition caused by this theory on just this issue. It also shows what kind of difficulties impeded this insight. It was these difficulties that discouraged Bohr and Einstein from understanding each other. In formulating their controversial theses, each of them had in view another aspect of the physical cognitive process, and thus were right in a certain

---

[2] Here classical mechanics is designated as protophysics in the sense that it is always necessary to go back to the original questions and fundamental concepts of this physical theory. This concept of protophysics must not be confused with that one developed by Hugo Dingler or with those formed by the constructivistic philosophy of science. For the concept of protophysics used here, see Refs. 41, 31, 52, 53; cf. also Ref. 51, especially pp. 64-81.

[3] Cited by W. Heisenberg in Ref. 21, p. 206.

[4] A. Einstein at the Fifth Solvay Conference, cited by W. Pauli in Ref. 27, p. 244.

way. For, as Bohr said: "The opposite of a correct statement is a false statement. But the opposite of a profound truth may well be another profound truth."[5]

2.1 On Bohr's and Einstein's Conceptions of Measurement and Reality

For Bohr, the characteristic feature of quantum mechanics is that one loses the possibility of distinguishing completely between the behaviour of the objects and their interaction with the measurement apparata determining the reference system [5]. Contrary to Einstein, he therefore considered, in accordance with Pauli and others, the quantum-mechanical state of a system as given only when the device by means of which it is measured is specified. Accordingly, Bohr accepted, at least for quantum mechanics, the need for necessity of certain measurement presumptions that determine what is meant in a given theory by such notions as "systems" and "state". He applied here the general rule: The real is that which can be measured.

Einstein, of course, accepted this rule. Nevertheless he did not subscribe to this definition of a quantum state since, as he argued, a sequence of numbers and measurement values cannot say what has been measured. The "what" is said by the theory.[6] And Einstein held the view that present quantum mechanics does not say sufficiently what one measures nor, accordingly, what is to be considered as real.

According to Bohr, the interaction between objects and measurement apparata represents an integral part of the description of quantum phenomena. Otherwise (and this was also pointed out by him), the measurement apparata must, in contrast to the quantum features of the phenomena, be described purely classically. Therefore, it was necessary to introduce a logical distinction between measurement apparata and atomic objects into the representation of experience [6]. Bohr concluded that quantum theory confronts us with a completely new situation, in which, notwithstanding the basic distinction of measurement apparatus and atomic object, subject and object cannot be uniquely separated.[7] Classical physical theories allow a unique separation, but they were, for him, idealisations, even though they have, of course, their own range of application. In reality, however, he believed, all is just like this as quantum mechanics says.

In regard to this epistemological interpretation, one can observe the following: In order to acquire physical knowledge, means of cognition are always required. Due to their mediating role, this means cannot be absolutely separated from either the subject nor from the object of cognition. They can and must, however, be uniquely distinguished conceptually from both. For each new theory, the means of cognition must be rediscussed and determined anew. The subject-independence of the objects investigated in the theory under consideration is thereby guaranteed. The difficulties in interpreting quantum mechanics are due to the problem of determining in them appropriate means of cognition (i.e., to clarify the measurement situation). This problem is complicated by the fact that

---

[5] Cited by W. Heisenberg in Ref. 21, p. 102.
[6] Cited by W. Heisenberg in Ref. 21, p. 77.
[7] Bohr thought that one finds such a situation also in psychology and biology, which leads to difficulties in these sciences.

quantum mechanics has to refer to classical mechanics in a twofold manner. Classical mechanics is not only a physical theory with a distinct range of validity, but it also forms an essential moment in the measurement basis presupposed for quantum mechanics. (Both are governed by the principle of correspondence and complementary.) The difficulties arising cannot, however, be explained by supposing a subject-dependence of quantum objects, where this dependence is welcomed as an expression of a new type of science.[8]

Of course, we do not want to ascribe this standpoint to Bohr completely. Bohr did not consider the role of measurement to indicate the indispensable necessity of a means of cognition mediating between object and subject of cognition. Rather, he considered his insight concerning measurement as a feature typical of quantum mechanics. But one should emphasise that: first, for quantum mechanics, Bohr reflected on the necessity of this means, and since he was convinced that nature behaves as described by quantum mechanics, this necessity had a general meaning for him; second, Bohr was opposed to a subjectivistic interpretation of this fact.

Bohr's position contains certain inconsistencies. On the one hand, he believed that quantum mechanics says how reality is constituted, while classical mechanics is only an idealization. On the other hand, he considered the relations and objects of classical mechanics as immediately resulting from our everyday experience [5]. This inconsistency resulted from the fact that he had in view only the connection between everyday experience and scientific experience, not their difference; he did not consider that classical mechanics is also based on scientific and not on everyday experience. This inconsistency and the apparent gap between classical and quantum mechanics thus have the same source. They result from an insufficient clarification of the epistemological status of measurement.

Bohr did not give a complete epistemological foundation of the quantum-mechanical features. Therefore, for some of his critics, his determination of the subject-object relation was not satisfactory. Indeed, Einstein found Bohr's objects to be "not sufficiently objective".

Einstein assumed a similar attitude as Bohr to experience, without accepting the conclusions drawn by Bohr from it [10, 11]. Roughly speaking, the Bohr's and Einstein's dispute did not derive from differing concepts of experience. Both believed that, in principle, one can compare a physical theory to reality without the interference of a bodily objective third,[9] at least via some (possibly complicated) conceptual steps. The discussions became controversial when they attempted to compensate this third (not completely conceived of or accepted by themselves) in a different way: Bohr accepted that one cannot abstract from measurement in quantum theory by assuming that one can no longer uniquely distinguish between the subject and the object of cognition. Einstein attempted to determine the object of cognition in a new manner, so that the means of measurement is not in evidence and thus no longer "perturbs".

Einstein's objection to the quantum-mechanical determination of a real is based on his consideration that, for physics, the possibility of isolated bodies or systems is essen-

---

[8] See Refs. 47, 49, 50; cf. also Ref. 51, especially pp. 82-108, and Ref. 60, especially pp. 74-95, 239-285.

[9] The term "bodily objective" denotes the philosophical concept known in German as "gegenständlich".

tial, both for distinguishing the subject and the object of cognition, and for distinguishing different objects. Einstein did not, however, consider the latter moment of this principle to be a general epistemological requirement, which had to be realised for each theory under consideration, but as a general physical requirement excluding actions at a distance. He wrote: "For the relative independence of spatially distant things (A and B) the following idea is typical: an external influence on A has no immediate influence on B. This is the proximity principle [or locality principle – R.W./H.-H.v.B.] which is only consequently applied in field theory. A complete repeal of this principle would make impossible the idea of the existence of (quasi-)closed systems and thus of the establishment of empirically probable laws in the current sense."[10, 11]

This argument presupposes that there is a physical reality in itself that is identical to objective reality, and that a quantum-mechanical description provides a certain probability for the appearance of an element of reality [34]. According to this conception of reality, the concept of a physical system is either given *a priori* or put by refined everyday experience (for Einstein, expressed by the principle of locality). Since, as Pauli stated, an isolated system is, however, defined by the theory under consideration (here, by quantum mechanics), Einstein's conception of reality must run into paradoxes [25].

This becomes especially evident in the famous EPR discussion, where Einstein tacitly works with two different conceptions of a system, namely with a quantum-mechanical one and with a space-time determination in the sense of the proximity or locality principle of classical field theory. After having used the knowledge from quantum mechanics that A and B form one system, Einstein continues to argue as if there existed two different systems in the sense of classical field theory or special relativity theory. This ambiguous definition of a system was criticised by Bohr and Pauli. As Pauli remarked [24], the ambiguity is related to the fact that Einstein, confused the terms "absence of interaction" and "independence" [43]. In essence, these critiques removed the EPR paradox.

The EPR discussion shows that Einstein founded the concept of system only by the locality principle, and, moreover, it makes clear that, for him, the isolated is realised only by an "atom", by an individual. Einstein accepted as genuinely real only individuals which, of course, could be complicated entities such as fields or formed from fields [9, 10]. Only an object in the sense of a thing in space-time was, in his view, something objective. Of course, an object determined in this physical way cannot be opposed to the subject. This opposition is, however, necessary and possible if one wants to discuss the concept of physical reality. The physical way of determining an object leads to the necessity of opposing objects only to their behaviour, and thus to the relation to one another. Apart from the above critical remark on the EPR paper, this feature is typical for

---

[10] See Ref. 11, pp. 321-322; cf. also Ref. 9.

[11] A similar position is taken today by many physicists. First of all, after the measurements by Aspect and his colleagues testing EPR-type predictions of standard quantum theory one finds the following standpoint: The phenomenon of quantum non-locality is experimentally confirmed; this does not mean an operational conflict between special relativity and quantum theory, yet there is a conflict with the spirit of a special relativity. Here the term "spirit of a special relativity" reveals the author's conviction that there is a reality in itself reflected by special relativity [1, 2, 28 (p. 245)]. – In parentheses, it should be mentioned that there are also physicists who do not accept the Aspect experiments as an evidence for quantum non-locality.

all definitions of "physical object" given by Einstein. This approach cannot be criticised; such is physics. But it involves the danger of transforming all epistemological (or philosophical) problems into physical ones such, that one forgets that one has used the specific physical approach and that there sometimes occur epistemological questions which must be discussed in these terms, and not transformed them into physical ones.

Einstein was thoroughly aware of the arguments that can be put forward against his reality concept. He summarised a typical objection as follows: "There we have the naked formulation of a metaphysical prejudice, empty of content, a prejudice, moreover, the conquest of which constitutes the major epistemological achievement of physicists within the last quarter-century. Has any man ever perceived a 'real physical situation'? How is it possible that a reasonable person could today still believe that he can refute our essential knowledge and understanding by drawing up such a bloodless ghost?"[12] And he answered: "'Being' is always something which is mentally constructed by us, that is, something which we freely posit (in the logical sense). The justification of such constructs does not lie in their derivation from what is given by the senses. Such a type of derivation (in the sense of logical deductibility) is nowhere to be had, not even in the domain of pre-scientific thinking. The justification of the constructs, which represent 'reality' for us, lies alone in their quality of making intelligible what is sensory given (the vague character of this expression is here forced by my striving for brevity)."[13] Einstein expressly remarked that thinking without the positing of categories and of concepts in general would be as impossible as is breathing in vacuum.[14] Accordingly, for Einstein, the real in physics is to be taken as a type of program to which we are not forced to cling *a priori.* But no one is likely to be inclined to attempt to give up this program within the realm of the "macroscopic". Now the "macroscopic" and the "microscopic" are so interrelated that it appears impracticable to give up this program in the "microscopic" alone.[15] However, due to the fact that Einstein assumes it to be clear *a priori* what must be presupposed categorically, this epistemological *Ansatz* does not completely take effect in Einstein's overall conception.[16]

Transforming epistemological principles into physical ones (in the sense of a *μετάβασις*) Einstein believed it was possible to justify a special conception of physics, namely a development of physics as unified field theory, by merely epistemological arguments alone. The fact that the particle notion of classical physics fails in quantum mechanics was not traced back to its epistemological ground but considered as an argument in favour of replacing the particle by the field notion. This conception, at the same time, implies the idea that the solution of the problems raised by quantum mechanics would require new physical quantities [10]. It should be stressed that Einstein's hope for such new quantities does not take the form of the so-called hidden parameters. His idea was rather to extend the space of experience determined measurement-theoretically, so that dynamics can be formulated in it.

---

[12] See Ref. 13, p. 667.
[13] See Ref. 13, p. 669.
[14] See Ref. 13, p. 674.
[15] See Ref. 13, p. 674.
[16] For more details on Einstein's reality concept cf. Ref. 23 and Ref. 32, pp. 53-63.

Einstein's attempt to solve epistemological problems by physical means follows his view that the physicist, and only the physicist, can clarify the epistemological basis of this science. For this purpose, he considered the theory of cognition as a necessary requirement. But he appears to have regarded it not as a discipline that one has to study, but as one resulting from everyday knowledge, so that the physicist, like each thinking man, is well versed in it [10].[17] Einstein did indeed emphasise that the physicist has to go beyond the borders of existing physics when he is looking for a new or a more solid basis of physics. His vision of the theory of cognition was determined by his view that science is nothing but refined everyday thinking.

The connection between the critique of concepts and the development of physics shown by Einstein also makes it clear that Bohr and Einstein were aiming at different points in their quantum-mechanics discussion. Einstein was looking for the basis of a unified physical theory (and found that it was not quantum mechanics). Bohr was discussing the epistemological basis of quantum mechanics and compared it to the basis of other physical theories, mainly to that of classical mechanics. Nevertheless, this does not mean that one can reduce their controversial discussion to a simple misunderstanding, because both were discussing the same topic insofar as they were talking about the world in itself. Their answers were, however, different. Bohr said that, in principle, all is constructed quantum-mechanically. Einstein thought that, in principle, physical objects possess a definite location and momentum – or other properties that can be ascribed to them as quantities.

When Einstein speaks about the foundations of physics, he does not consider the epistemological or measurement-theoretical basis of physics. He was not asking for the means of physical cognition, but for that physical theory which could provide the basis for unifying physics. This is why he considered classical physics to be at the centre of physics. In its relativistic version, classical mechanics was for him the dynamics underlying all physics [10].

All in all, on the differences in Bohr's and Einstein's conception of physics, one may state the following: Einstein tried to make the means of cognition (not identified by him as that) the object of cognition or, put otherwise, to unite them, while delimiting the subject of cognition uniquely from both. Bohr wanted to include the means of cognition (not identified by him as such, either, but considered as part of the subject of cognition) in the object of cognition in such a way that there was no unique distinction between subject and object.

Einstein's uneasiness with quantum mechanics is thus due to the fact that this physical theory does not permit one to delimit clearly the object/means of cognition from the subject of cognition. Similarly Bohr, who was influenced, like all his contemporaries, by the mechanistic interpretation of mechanics,[18] was initially shocked by the fact that quantum mechanics does not allow the unique separation of the subject and object of

---

[17] Of course, a philosopher who is not well versed in physics cannot say anything on the epistemological status of physics or on its philosophical problems.

[18] For the reasons underlying mechanistic interpretation of the classical mechanics and the difference between the mechanics and the mechanistic world picture, cf. Refs. 46, 54, 61.

cognition which he believed to exist in classical mechanics. Subsequently, he interpreted just this feature of quantum mechanics as expressing its greater realism.

2.2 The Subject of Cognition in the Scientific World Picture

Many problems concerning the concept of realism are thus connected to the difficulty of thinking the means of cognition that mediates between subject and object of cognition, and this difficulty is caused mainly by the concept of subject that one has to presuppose in physics.

The typical feature of the physical concept of subject was especially clearly worked out by Schrödinger. Schrödinger's concept of subject was intimately connected to the epistemological ideas on which his wave mechanics was founded, and to those which followed from it. Schrödinger's labours may be summed up as an answer to the question of how continuity and discontinuity are related. As a pupil of Boltzmann's, he was fully convinced of the profound significance in physics of the Boltzmann's atomism – both, that is, of Boltzmann's thermodynamics and of Boltzmann's gnoseological point of view. It was in part such ideas which led Schrödinger to develop wave mechanics.[19] But once he realised that it was an illusion to believe that wave mechanics could be interpreted as a pure continuum conception, he began to study the epistemological foundations of science.

True, the founders of wave mechanics "indulged for some time in the fond hope that they had paved the way of return to a classical continuous description of nature", as Schrödinger conceded in a talk on "Science and Humanism".[20] But he was not content simply to accept that this hope had failed, and he began to search for the reason for its failure. He found that it was caused by the problem of grasping the continuum conceptually. In his view one could not simply reject the possibility "that our present difficulties in physical science are bound up with the notorious conceptual intricacy inherent in the idea of the *continuum*".[21] We mostly overlook this possibility, because we consider that we are familiar with the concept of a continuum and know to handle it.[22] Ancient thinkers, on the other hand, knew that the concept of continuum has its problems. Indeed, they invented atomism in order to overcome them. This invention was outstanding enough to carry the weight of science to the present day; and Schrödinger sings the praise of atoms and quanta as the age-old counter-spell against the magic of the continuum.[23]

It may seem surprising to call what forms part of the foundation of modern science an "age-old counter-spell". But does this surprise not have its basis in our lack of surprise that the world is comprehensible, that motion can be circumscribed by laws, i.e., by stable relations? The comprehensibility of the world is so self-evident for the modern scientist that it does not even occur to him ask what assumptions underlie it. Yet just this is

---

[19] For details, see Ref. 48.
[20] See Ref. 36, p. 158.
[21] See Ref. 36, pp. 157-158.
[22] See Ref. 37, p. 62.
[23] See Ref. 36, 38; cf. also Ref. 37, pp. 3-21, 90-98.

the question that in Schrödinger's view must be put if the difficulties of modern physics are to be understood and hence also solved.

The ancient atomistic method of conceiving the continuum by way of the discrete is rooted, according to Schrödinger, in two principles: on the one hand, the assumption that what happens in the world may be comprehended, and, on the other, the fundamental rule of eliminating the recognizing subject from the desired image of the world and letting it play the role of an external observer.[24]

The first principle (principle of comprehensibility) found its justification when Ionian natural philosophy overcame the mythological way of thinking. This ascendancy initiated the transition to a rational view of the world and thus provided a basis both for philosophy and for science.[25]

The second principle (principle of objectivisation) arose from atomism, which regarded the entire world as made up exclusively of atoms and the vacuum. Thus the subject of cognition was seen as a mere natural object and hence eliminated *qua* subject. But when ancient atomism attempted also to explain the soul as built up of atoms, it entered into contradiction with its own fundamental principle and gave rise to paradoxes.[26] Therefore, it is important to say: The atomist's solution for making motion thinkable and bringing thought and reality into coincidence was one possible solution of this problem of motion, thought, and reality. It was that particular solution which took nature as its undoubted presupposition. This is just the point of view of (natural) science. The other possible (and indeed necessary) solution to the problem is the philosophical one founded all in all by Plato. The distinction between the two solutions as regards the atomistic thesis may be simply stated: Philosophical atoms cannot exist, but physical ones must exist.[27] Of course, the ancient atomism was to be modified by Newton epistemologically to form an important principle of physics that has not needed revision to the present day. – Schrödinger considered that omitting the recognizing subject is an artifice, yet one that only a fool would not make use of.[28] But he took Ionian natural philosophy to have done no more than create a precondition for scientific rationality, and only traced out the later development of this rationality; hence he simply failed to see the recognising subject in the modern picture of the world – the subject from which abstraction is made in fact only in the sciences, not in philosophy.

Schrödinger's analysis of what ancient atomism means for modern physics leads to the following conclusion: However much the concept of an atom may have changed (the atoms of quantum mechanics are no longer individuals whose motion we may follow through time) we cannot do without this concept; taking atomism here not so much in the sense of the theory that matter is composed of minimal particles but in the sense of an "artifice" by means of which the conceptual difficulties of the continuum may be mastered [38]. Here Schrödinger terms a thought-principle which is epistemologically founded an "artifice". He describes it as a necessary tool that can deform our picture of

[24] See Ref. 39, pp. 126-137, and Ref. 37, pp. 90-98.
[25] See Ref. 56, pp. 12-31.
[26] See Ref. 37, pp. 75-89.
[27] On atomism as thought principle, i.e., as the thought principle of science but not of philosophy, cf. Ref. 57.
[28] See Ref. 37, p. 93.

the world if we forget that we have used it and if we do not know what it consists in. But such a deformation occurs only in these cases. (Schrödinger's concept of atom as a thought-principle differs strongly from the understanding of atom as a thing in space-time.)

According to the analysis given by Schrödinger, the removal of the cognising subject from the world-picture is a necessary consequence of the hypothesis of comprehensibility.[29] It is equivalent to setting the hypothesis of a real outside world. It is, as he stated, a convention comparable to that by which the citizens of a town agree to put private interest aside in favour of the public welfare [35].

In so arguing, Schrödinger determined the main feature of the epistemological starting point of natural science. For natural science, objectivity grasped in contrast to subjectivity is a sign of true cognition. The world is grasped by science in the form of the object. Therefore, the subject as such cannot be considered from this point of view. It remains as an exterior observer and it is reflected implicitly in the thesis that an objective outside world is given. – To consider the subject *as* subject is the task of philosophy.

Here a general remark on the relationship between philosophy and (natural) sciences is in order. The truth of the matter is not only that philosophy is an important prerequisite of the sciences, but also that the empirical sciences, in particular physics, are among the major prequisites of philosophy. Natural sciences and philosophy need one another. It is not the case that philosophy might ever be able to dispense with the sciences, for they must always remain one of its premises.

When the significance of philosophy for physics is stressed here, this does not mean that philosophy might or should make proposals for a reformulation of physics. As its relation to physics, the task of philosophy is to clarify the origin of the thought-determinations employed by physics. This clarification means recognising of the categories being employed. The aim is to ascertain the epistemological status of physics so as to answer the old Kantian question of *how science is possible*. This will certainly involve the investigating the basic characteristics of the experimental method and of *scientific* experience as such. And this much is clear: the natural or physical sciences themselves realise their own "thinking attitude to objectivity" or, in Hegelian terms, their own "position of the thought to objectivity". In other words, the position of the thought to objectivity realised by the sciences is identical neither to that of empiricism nor to that one of rationalism.

2.3 Means of Cognition and the Physical Concept of Space

In a physical theory, the subject of cognition is deputising by the measurement means, so that the subject occurs in an objectivised form. Schrödinger did not draw this in this form. Although claiming that man was not replaced by the means of measurement step by step, but was so from the very beginning on, Schrödinger interpreted this insight in the way that the subject is eliminated and not objectivised. This is due to the fact that he con-

---

[29] See Ref. 35; cf. also Ref 39, p. 129 and Ref. 37, p. 92.

ceived of the subject as a human individual, and not of the subject as the human species, he conceived of the subject as the thinking human individual and not of the subject as the intellectually and objectively (*gegenständlich*) working human species.

In quantum mechanics, it became evident that one cannot speak of systems or states as such in physics. All these entities are determined only when the measurement apparatus by which they are specified is given. Conversely, these measurement apparata may only be determined with respect to quantum mechanics. Thus the term "reality as such" has no sense.

More technically, the role of measurement is revealed in quantum mechanics by the fact that the space of measurement and the space of the representation of physical dynamics do not coincide. The measurement space is the three-dimensional (or four-dimensional) space of experience, in which the measurement apparata are defined, while the spaces in which the physical basis laws are formulated are higher-dimensional configuration or phase spaces. To establish a connection between measurement and dynamics, supplementary axioms concerning measurement must be formulated. These axioms say how one can transfer from the high-dimensional representation-space of dynamics to the four-dimensional space-time of measurement (the transition is, of course, not "distortion-free" in general). And since Einstein and others did not want to accept this noticeable role of measurement imprinted on the theory, they tried to formulate a geometrically unified theory where only one (albeit more complicated) geometrical space-time arises. Other authors look for an improved quantum theory either eliminating EPR-like situations or, if one is prepared to admit them, at least eliminating the reduction of the state vector.

To repeat, in quantum mechanics the role of measurement becomes especially evident. Our thesis reads: Part of the shock that one finds in studying quantum mechanics results from the fact that the mechanistic interpretation of classical mechanics hides that one finds there a similar state of affairs. In particular, classical mechanics also requires a dualism between the representation space necessary to formulate dynamics and the measurement space which is an essential moment of the physical means of cognition.

For classical mechanics is concerned with measurable quantities and not with pure mathematical entities, or sensual-concrete individuals. This means that classical mechanics, like quantum mechanics, needs a space (or space-time) that is not given neither as a mathematical space, nor as the space of our everyday experience. It is a space determined by the requirements of measurement. Furthermore, the physical laws are relations between these measurable quantities and quantities that are at least somehow related to the measurable ones. To formulate all the laws and relations of a physical theory, classical mechanics requires mathematical auxiliary quantities and corresponding relations between them; thus one needs a general mathematical representation space. It was just Helmholtz' and Hertz' papers that demonstrated that one should use such general spaces to extend mechanics to its extreme range of validity. The fact that in classical mechanics (and field theory) one *can* formulate the dynamical relations in an Euclidean space (or a Minkowski space) has concealed the space dualism that exists in classical mechanics as well.

Knowing only classical mechanics, it is of course difficult to appreciate the importance of the abstract spaces used in dynamical representations. Their use seems no more than a mathematical trick fruitful for practical applications of mechanics. Today it is, however, noticeable that this reflects the fact that physics needs, in principle, two spaces – the measurement space and the space in which dynamics is formulated – where both spaces are theoretical constructions and thereby differ from the space of everyday experience. For each physical theory one has to clarify the relation between these two spaces (otherwise the "dynamical laws" are pure mathematical formulas and not physical laws). In classical mechanics, this relation is given mainly by the following requirement: Find a representation of the dynamics that needs only Euclidean space notions. In quantum mechanics such a representation does not exist. Heisenberg was talking therefore even of a cut between the two spaces. He said: "The situation of any experiment in nuclear physics is as follows: We use more or less complicated apparata for putting questions to nature, questions that are always directed to the definition of any objective process in space and time … . Thus it is an automatic result that in dealing with the process mathematically, we make a cut between the apparata we use as auxiliary means for putting our questions, on the one hand, … and the physical systems about which we want to learn something, on the other. We the latter represent mathematically by a wave function and for this wave function a differential equation exists according to the quantum theory which determines the future state of the wave function on the basis of the present one. As to the apparata, however, we content ourselves with laws formulated in classical concepts which provided us with the right to use them as measurement-apparata. The cut between the system to be observed and the measurement apparata is given by our questions naturally, but it does not show a discontinuity of the physical process obviously."[30]

But because, for reasons of measurement, one needs a reference to Euclidean space (and thus to classical mechanics), one must also overcome this cut. This is done by certain projections from the representation onto the measurement space.

In a lecture given in 1932, Max Planck expressed this situation as follows: "Generally, the laws of material waves are completely different from those inherent in the classical mechanics of material points. What matters, however, is the fact that the function characteristic for material waves, the wave function or probability function (it does not matter which name is used), is completely determined by the initial and boundary conditions for all places and times, according to certain rules of calculation, whether one uses *Schrödinger* operators, *Heisenberg* matrixes, or *Dirac* q-numbers. – We see: in the world picture of quantum mechanics determinism governs just as it did in classical physics, it is only the symbols and stipulations of calculation that are different. According to this, in quantum physics, as in classical physics before, the uncertainty of predictions relating to

---

[30] "Die Situation irgendeines Experiments in der Atomphysik ist die folgende: Wir stellen mit Hilfe von mehr oder weniger komplizierten Apparaten Fragen an die Natur, die stets auf die Feststellung irgendeines objektiven Ablaufs in Raum und Zeit gerichtet sind … . Aus dieser Sachlage ergibt sich automatisch, daß wir bei der mathematischen Behandlung des Vorganges einen Schnitt ziehen einerseits zwischen den Apparaten, die wir als Hilfsmittel zum Stellen der Frage betrachten, … und andererseits den physikalischen Systemen, über die wir etwas wissen wollen. Die letzteren stellen wir mathematisch durch eine Wellenfunktion dar, wobei für diese Wellenfunktion nach der Quantentheorie eine Differentialgeichung besteht, die den zukünftigen Zustand der Wellenfunktion aus dem gegenwärtigen bestimmt. Für die Apparate dagegen begnügen wir uns mit den in den klassischen Begriffen formulierbaren Gesetzen, die uns das Recht geben, sie eben als Meßapparate zu verwenden. Der Schnitt zwischen dem zu beobachtenden System und den Meßapparaten ist durch unsere Fragestellung naturgemäß gegeben, bezeichnet aber offenbar keine Diskontinuität des physikalischen Geschehens" [20].

events in the sensual world is reduced to the uncertainty of the connection between world picture and sensual world, that is, to the uncertainty of translation of world-picture symbols to sensual world, and vice versa."[31]

The circumstance that the basis of quantum-mechanical measurement is more complicated than that of classical mechanics should not be interpreted as a "loss of reality". It only appears so if one believes that refined everyday experience determines Euclidean space (or Minkowski space) as *the* real space (or space-time), and if one does not recognise this space as a theoretical construction required for measurement and forming one part of a dualistic space conception. From this point of view, Einstein's grand program of a unified field theory, mentioned above, can be interpreted as an attempt to weaken Heisenberg's cut in such a manner that jumps do not arise – in other words, to find a less dualistic physical theory. Such a unified theory would represent a great step forward, but it would be difficult to justify its necessity by looking to seeming paradoxes of quantum mechanics or by appealing to everyday experience.

Therefore, in its essence, the discussion on quantum mechanics centres on the new construction of a means of cognition and on the consciousness that such a means is *always* necessary. Of course, here one also is concerned with the new construction of the physical object and the epistemological status of the physical object at all. This discussion is frequently not seen in those terms because, in principle, it involves an confusion or an identification of means and object. When the means of cognition is discussed here, the *physical* means of cognition is always meant. This is the space-time structure presumed in each case, and its objectivisation (*Vergegenständlichung*) in the ideal constructions of the experiment. The physical objects are grasped as dynamical quantities. (The physical objects are not the sensual-concrete things.) In this manner, the physical quantity is fixed as the object by physicists; however, considered epistemologically, the physical quantity being a thought-entity is a means. A physical quantity is, from the epistemological point of view, a means, and, from the physical point of view, an object. Physics as physics is right when it proclaims this object to be the object at all. However, when epistemological discussions are the point, then this identification of the physical quantity must be abrogated. If one sees only the first aspect of the quantity (the means aspect), one is led to subjectivistic interpretations; if one sees only the second one (the object aspect), one is led to an ontologistical interpretation of physics. In both cases, however, one makes an effort to rereduce the three momenta of the epistemological pro-

---

[31] "Im allgemeinen sind die Gesetze der Materiewellen grundverschieden von denen der klassischen Mechanik materieller Punkte. Wesentlich ist aber der Umstand, daß die für die Materiewellen charakteristische Funktion: die Wellenfunktion oder Wahrscheinlichkeitsfunktion – der Name tut hier nichts zur Sache – durch die Anfangsbedingungen und die Randbedingungen für alle Orte und Zeiten vollständig determiniert ist, nach ganz bestimmten Rechenregeln, sei es, daß man sich dabei der *Schrödinger*schen Operatoren oder der *Heisenberg*schen Matrizen oder der *Dirac*schen q-Zahlen bedient. – Wir sehen: In dem Weltbild der Quantenphysik herrscht der Determinismus ebenso wie in der klassischen Physik, nur sind die benutzten Symbole andere,und es wird mit anderen Rechnungsvorschriften operiert. Dementsprechend wird in der Quantenphysik, ebenso wie früher in der klassischen Physik, die Unsicherheit in der Voraussage von Ereignissen der Sinnenwelt reduziert auf die Unsicherheit des Zusammenhangs zwischen Weltbild und Sinnenwelt, d.h. auf die Unsicherheit der Übertragung der Symbole des Weltbildes auf die Sinnenwelt, und umgekehrt" [29]. – The fact has to be taken into account that Planck uses the term "sensuous world" instead of the terms "space of physical experience" or "space-time of measurement".

cess (object, subject, and means of cognition) to two. This abstraction from the means keeps one from recognising and specifying the *real-general.*

It follows from this treatment that the concept of physical reality discussed here from the viewpoint of measurement does not support a positivistic interpretation of quantum mechanics (or physics at all). Rather it means of rediscovering Kant's Copernican turn.

In seeking an answer to the question how necessary and universally valid statements on reality are possible, Kant realised that a mediating instance is required if mind and sensibility (or theory and reality) are to connect. He grasped space and time as being *a priori* given pure forms of sensibility for this mediating instance [22]. Here Kant reflected in a general philosophical way the function which Newton ascribed to geometry in the framework of classical mechanics. In the Preface to his "Principia" Newton wrote that geometry lies at the basis of practical mechanics and represents at the same time that part of general mechanics which establishes and proves the art of exact measurement. Clearly, that Kant's space-time apriorism must be criticised [44, 58]. Kant's insight, however, that the transition from ideality to reality cannot be carried out without a mediating third can not be overturned.[32]

The long discussion on the epistemological status of quantum mechanics does not only show that physics (like other sciences) requires a means of cognition that must be thought together with the special physical theory under consideration. It shows also that for each new theory these means must be rediscussed and freshly determined. Finally, it shows that physics does not speak of reality in itself but of physical reality depending on the physical theory considered, including its means.

3. Heisenberg's Cut and the Limitations of its Movability

Since the founding of quantum theory, many authors have considered the physical reality concept and the role of measurement. As said above, such discussions follow papers of the founders of quantum theory, and the difficulties with which they grapple are already known from those classic papers, too. One of these difficulties was clearly expressed in the papers of Bohr and Heisenberg on the uncertainty relations arising in quantum electrodynamics. Their conclusions were based on different views of the movability of the so-called Heisenberg cut and, in this context, of the requirements of measurement. As a contribution to current discussions, these different views shall be worked out in the following. In this connection, the restrictions on the movability of the Heisenberg cut are demonstrated (Section 3.2). Before doing that, we shall give arguments in favour of the occurrence of the cut and its movability (Section 3.1).

[32] For details, see Ref. 44, 58, 59.

3.1 Heisenberg Cut

According to W. Heisenberg, the quantum-mechanical description starts in discord [19]. Heisenberg characterises this discord by stating that, on the one hand, the task of physics consists in giving a lawful description of objective processes in space and time, whereas on the other hand, for the mathematical representation, it uses a ψ-function in a configuration space that one cannot simply consider as a representation of objective processes in space and time.[33]

This conflict is manifested in an arbitrariness of the application of quantum mechanics [19]. It remained an open question as to whether only the observed atomic system should be described by a wave function, while the apparata used for the observation should be described according to the laws of classical mechanics, or whether these apparata should also be represented by a wave function according to the laws of quantum mechanics. (In the latter case, according to Heisenberg, one had to consider only the observation of the measurement apparatus, e.g., the observation of a curve on a photograph as a "classical-illustrative" event.) Heisenberg expressed this arbitrariness by posing the question: "Where should one make the cut between the description by the wave function and the classical-illustrative description?" [19]. And he arrived at the conclusion that the quantum-mechanical statements on the results of experiments do not depend on the position of the cut. More precisely: The cut can be moved arbitrarily far in the direction of the observer, up to the region that usually is described according to the laws of classical mechanics; however, it cannot be moved arbitrarily in the direction of the atomic system, since there exist physical systems (and all atomic systems belong to them) for whose description the classical concepts are inappropriate, and whose behaviour thus can be expressed only in the language of the wave function.[34]

In giving the reasons for the occurrence of the cut, Heisenberg referred mainly to the necessarily final "observation". He argued as follows [18]. The system A which is to be investigated is described by the Schrödinger equation. The measurement values of the quantities which are connected by Schrödinger's equation, cannot, however, be predicted – only the probability of finding a definite value in a single measurement. This blurredness is caused by an uncertain perturbation which is produced by the influence of the measurement apparatus B. Now one could of course also describe the system B quantum-mechanically. Then the influence of the "measurement apparatus" on the system A becomes determinate; it could be described by a Schrödinger equation with the Hamilton operator

$$H = H_A + H_B + H_{AB} \tag{1}$$

where H_{AB} gives the interaction between A and B. But in this case the measurement apparatus becomes a part of the quantum-mechanical system, so that another system C is required that observes the system $A+B$. This observation leads one to the uncertainty

---

[33] Schrödinger described the situation similarly [33]. He stated: Quantum mechanics regards classical mechanics as suited to inform us about those measurements in space and time which in principle can be performed on the corresponding natural objects, but it declares classical mechanics incompetent for describing the connection between the physical quantities.

[34] For a discussion of Heisenberg's cut, cf. also Refs. 40, 42.

one wished to avoid by adding B to A. For, according to Heisenberg, the notion "observation" belongs to the world of everyday experience and one always has a complementarity between this world and the causal description by the Schrödinger equation. This complementarity is expressed in the occurrence of the cut between the system whose quantities are to be measured ("the observed system") and the measurement system ("the observing system").

For Heisenberg, the observing system can be a photograph, or even the human eye, because – as he saw it – the final aim of physics is to describe experiments and their results just as one does the things and events in everyday life, i.e., by concepts concerning the space-time world obviously surrounding us, which concepts are expressed in the usual language appropriate to this space-time world.

Even though the occurrence of the cut is proven by the need to treat finally one of the systems A, B, C, ... as the observing system, this still does not contradict the above statement concerning arbitrariness. This arbitrariness, which leaves physicist free to decide which objects he considers to be part of the observed system, and which he regards as the observational means, can of course only occur if it does not get into contradictions with the experimental predictions calculated for A. Thus Heisenberg showed that the measurement system B can be described quantum-mechanically (together with the observed system A) without contradiction [19].

His considerations make clear that the mutual influence of the parts of a system on one another must be described differently from the influence of the measurement system on the system to be measured. In the first case, the influence is accounted for an interaction term H_{AB} in the Hamiltonian; in the second case, the coupling of the measurement apparatus and the system A leads to the probabilistics of quantum mechanics and to interference phenomena which cannot be reduced to interaction.

The latter point was stressed by Pauli [24] when he stated that it was the mixture of the notions "absence of interaction" and "independence of the two parts A and B of the quantum-mechanically described system $A+B$" that led to the paradox formulated by Einstein, Podolsky and Rosen [15]. Even if there is no interaction between two particles A and B, they need not be independent of each other. Considering them as forming one quantum-mechanical system, they are described by one ψ function interfering with itself and only with itself. As a result, there can be exist an entanglement (Schrödinger) between A and B such that their coordinates $q_A \cdot q_B$ and momenta p_A and p_B can obey the relation

$$q_A = q_B,\ p_A = -p_B. \tag{2}$$

Accordingly a measurement on A or B is automatically a measurement on the whole system $A+B$. If one now measures, e.g., q_A then one may predict q_B. But one then destroys the entanglement of A on B and cannot use the relation (2) for further measurements. Therefore, a paradox does not occur as long as one argues within the framework of quantum mechanics. If, however, one considers the two entangled parts A and B under the viewpoint of interacting and non-interacting systems (as was done in the EPR paper) then one follows a line of argument that conflicts with quantum mechanics.[35]

---

[35] For these and further arguments, see Ref. 43.

Therefore, the EPR effects (or the phenomena of quantum non-locality) do not represent a paradox. A paradox appears only if one presupposes features of reality that are external to quantum theory here under consideration.

A similar position was taken by Heisenberg [19]. His considerations concerning the movability of the cut aimed at showing that a "deterministic completion" of quantum mechanics cancelling its probabilistic character cannot exist, at least as long as one does not consider the most essential experimental successes as accidental ones. From the viewpoint of a distinction of interference and interaction, Heisenberg's argument using the movability of the cut can be described as follows: Let us consider an α particle emitted by a radioactive nucleus. According to quantum mechanics, one can make only statistic predictions for the time of the decay and for the direction of emission. But one knows that, if the α particle is diffracted by a grating, one finds a diffraction pattern on a screen standing behind it. Assuming now that there was a completion of quantum mechanics which enabled one to predict the moment of decay and the direction of the emitted particles, it would then follow that the experimental (diffraction) results of quantum mechanics could not be calculated. Indeed, if we were to move the cut so that the grating is also described by quantum mechanics we could, due to our improved knowledge, consider the interaction between the α particle and the grating. In this case, the influence of the grating on the particle would depend only on the point of impact of the particle, and not on the structure of the whole grating. Consequently, one would not find an interference of the ψ-function and thus no probability behaviour.

The argument leads to the conclusion that the experimental results of quantum mechanics cannot be reproduced if one completes quantum theory so that interaction (in quantum mechanics belonging to the particle aspect and regarded in the Hamiltonian) plays a greater role. One cannot do that without getting in trouble with the experiments confirming the wave aspect of quantum mechanics. The same happens if one tries to complete quantum mechanics by bringing more wave aspects into the theory. Indeed, one then runs the danger of contradicting those experiments which reveal the particle aspect. Therefore, theories working with hidden parameters and theories aiming at a more realistic interpretation of quantum mechanics often conflict with experiments or result in a new theory.[36]

As was mentioned at the outset, Heisenberg considered the disagreement between the Schrödinger equation formulated in abstract mathematical spaces and the classical-illustrative measurements in the space-time world to be a problem. Summarising the above, one can say that, in discussing this problem, he introduced the thought that there is a principal difference between the coupling of an observed system A to the observing system B and the coupling between two parts A and B of a single system described quantum-mechanically. In the first case, a cut arises. This cut can be moved by describing B (or parts of B) quantum-mechanically as well. However, this circumstance which results

[36] The linearity of Schrödinger's equation requires to interpret $|\psi|^2$ as probability. This leads to effects which can only be grasped from the point of view of interference and not of interaction. A. Barut shows that there are arguments in favour of an equivalence of this linear theory to a non-linear theory, where $e|\psi|^2$ can be interpreted as charge density and where the effects are explained by interaction instead by interference mechanism [4]. Such a non-linear theory is an example of a new theory.

for reasons of self-consistency of quantum mechanics does not alter the fact that B, if considered under the aspect of measurement apparatus, has to be connected with A in a fundamental different manner.

It is evident, the determination of the Heisenberg cut implies the statement that the cut cannot be arbitrarily far moved in direction of the quantum system on which measurements are to be performed. But Heisenberg believed that, as far as the movability in direction of the experimenter is concerned, one has great latitude. As mentioned above, this opinion goes back to his demonstration that the measured values calculated for the system A described quantum-mechanically do not change when one adds B to A and performs the calculations for $A+B$, while the measurements are done by means of an apparatus C. Heisenberg was thereby led to the conclusion that the cut, although unavoidable, should in principle be movable to the position just in front of the eye of the experimenter. According to Wigner [63], the "dividing line" between observer and observed physical object can even be shifted towards the mind of the observer.

This latter conclusion, however, is very problematic, as can be seen by comparing Heisenberg's arguments with those Bohr and Rosenfeld gave in the framework of investigations on quantum electrodynamics.

3.2 Limits to the Movability of the Cut

Bohr's and Rosenfeld's arguments [7] do not immediately concern quantum mechanics but quantum electrodynamics; however, they can easily be compared with those of Heisenberg because he himself transferred the quantum-mechanical considerations discussed above to quantum electrodynamics [18].

Referring to the δ-function in the commutation rules for the components of the electromagnetic field, Bohr and Rosenfeld started with the statement that quantum-theoretical field quantities like the x-component of the electric field are not genuine point functions but space-time integrals of the type

$$\overline{\mathfrak{E}_x} = \frac{1}{VT}\int_T dt \int_V \mathfrak{E}_x dv \qquad (3)$$

(V denotes the three-dimensional volume and T the corresponding time interval). For our discussion the requirements on the measurement of such average field components formulated by Bohr and Rosenfeld play an important role. Therefore, let us here briefly summarise the requirements in the case that a field measurement is performed by two different measurements of the momentum of a test body:[37] (i) In order to perform optimal field measurements, one must have the charge of the measurement body at one's disposal. This means, in particular, that one has to realise a measurement body whose atomistic structure can be neglected. (ii) If one assumes that the two momentum measurements are performed on a T interval, the duration Δt of each measurement has to be much smaller than T. Only then can the electromagnetic backreaction of the accelerated measurement body be neglected. (iii) To measure effects of quantum electrodynamics one has to as-

[37] For more details, cf. also Ref. 45.

sume $L > cT$ (where $V \sim L^3$). Then one measures in a region where one can test characteristic consequences of the quantum formalism without running the risk that the effects are drawn in quantum fluctuations. This means, of course, that one has to realise a measurement body which behaves as a non-relativistic rigid body.

Generalising these and other requirements on an optimal measurement, Bohr and Rosenfeld claimed that the test bodies which are to be used in quantum electrodynamics have to behave as rigid bodies with a homogeneously distributed charge according to the rules of classical mechanics and electrostatics. In this context, they argued that, in all measurements of physical quantities, it must by definition be a matter of applying classical presentations. This is also true for field measurements. Any objection to this, referring to the limited range of validity of classical theories, would be in contradiction to the concept of measurement. As far as the displacement of the test body is concerned, however one has to regard quantum mechanics; otherwise one would make a classical instead of a quantum measurement.

Under realising these conditions, one can show that the measurement possibilities agree with the complementary limitations on measurability resulting from the formalism of quantum electrodynamics. In their paper Bohr and Rosenfeld showed in particular, that the average values of all field components over the same space-time region commute; they are thus exactly measurable independently of each other.

In using the above-given arguments, Bohr and Rosenfeld criticised results derived by Landau and Peierls, as well as by Heisenberg.[38] In particular, in his work "The Physical Principles of Quantum Theory" Heisenberg had concluded that, while the average values of single field components can be measured with an arbitrarily high accuracy, for different field components, e.g., for $\bar{\mathfrak{E}}_x$ and $\bar{\mathfrak{H}}_z$, one has

$$\Delta\mathfrak{E}_x \Delta\mathfrak{H}_z \geq \frac{hc}{(\delta l)^4} \tag{4}$$

(δl denotes the spatial extent of the measurement apparatus or of the domain over which the average is calculated). Bohr and Rosenfeld showed that the uncertainty relation (4) results when one does not consider optimal measurements, i.e., measurements that do not satisfy the above conditions. In Heisenberg's paper, these conditions were violated by using electrons (or point charges in the sense of quantum or atomic theory) instead of classically structured test bodies.

This insight into the general nature of measurement gained by Bohr and Rosenfeld in the context of rather special considerations on quantum electrodynamics makes clear one can neither assume an arbitrary movability of Heisenberg's cut in the direction of the observer nor characterise a classical-physical or classical-mechanical measurement apparatus as a mere illustration or visualisation of physical results.

The fundamental statement that the cut must be movable does not entail arbitrary movability. Not only there are limitations in direction of the atomic system (the object of cognition), which were emphasised by Heisenberg, but also there are limitations in direction of the observer. If one ignores these latter limitations on the movability of the cut by

[38] For the discussion between Landau/Peirls and Bohr/Rosenfeld and for the implications of the given arguments for quantum gravity see Ref. 45.

allowing the use of a measurement apparatus that is partly a quantum or atomic system (which cannot be described as a classical physical body), one does not perform an optimal measurement.

When we speak of the advantage that, in some respect, Bohr's and Rosenfeld's characterisation of measurement has over Heisenberg's we do not thereby intend to deny that Heisenberg also considered the cut to be unavoidable.[39] For, if one were to treat the measuring apparatus as a part of the quantum system, one would need a new measuring apparatus in order to observe the extended quantum system. For this apparatus, however, the same considerations would apply as those given above, and, following Heisenberg, one should be forced to include this system and so on – up to the point where the whole universe would be considered as a single system – "but then physics has vanished, and only a mathematical scheme remains".[40]

Due to these arguments which are of great epistemological value, alternative interpretations such as those of Everett and others [3, 16, 62] are most problematical. For they seek an interpretation whose fundamental ingredients do not include the concept of a measurement by an "external" apparatus or observer.

In this context, it is especially interesting to cast a short look at a paper by Cini [8], where an attempt to overcome the cut can be reinterpreted in favour of the cut's necessity.

Cini, following Schrödinger and other authors, starts with the assumption that not only the (quantum) microsystem to which the measured variable belongs, but also the apparatus devised to perform this measurement have to be treated quantum-mechanically. If one does so, the measurement apparatus loses the double nature which leads to absurdities. For example, it would seem that one is entitled to treat the interaction of a microsystem with a given macroscopic object by means of the Schrödinger equation, so long as one does not know the object is a measuring apparatus; conversely, as soon as one takes into account its role as a measurement device, one is be forced to describe it as an intrinsically unanalysable object not subject to the laws of quantum mechanics. By describing microsystem + apparatus quantum-mechanically, the postulate of the wave packet collapse or of the projection of the wave function should be eliminated and replaced by Schrödinger time evolution of the state vector of the total system microsystem + apparatus. This total quantum system should establish a one-to-one correspondence between one of the physical variables or states of the apparatus and the physical variable of the microsystem brought into interaction with the apparatus. As a simple case, Cini shows that, if the number of the particles making up the measuring apparatus is large, one does find such a one-to-one correlation. As a consequence, for Cini the wave function collapse is only an approximation for describing what happens to the microsystem on which a measurement is performed: All can be calculated if one takes the quantum structure of the measuring apparatus into consideration.

Thus, both the collapse and the Heisenberg cut would seem to be eliminated. Cini emphasises that the cut only occurs when the apparatus is assumed to be made of an in-

---

[39] In the German 1942 edition of the work "The Physical Principles of the Quantum Theory", Heisenberg refered to the 1933 Bohr-Rosenfeld paper, but without discussing in detail.

[40] See Ref. 18, p. 58.

finite numbers of particles. He takes this to corroborate his interpretation of measurement procedure. For, "since any physical instrument is made of a finite number of particles, the collapse is only an approximate description of the actual time evolution of the system (object + apparatus)".[41]

The considerations outlined above, in our view, are convincing in two respects: first, they show the necessarily microscopic character of the measuring apparatus; second, they take only the measuring apparatus and its interaction with the microsystem as an ingredient of the discussion, but not the sensation and consciousness of the observer. The latter is presupposed actually but considered as lying beyond the analysis of the measuring process.

To our mind, however, such investigations cannot really eliminate the Heisenberg cut. They try to do so by assuming the seemingly natural postulate that the measuring apparatus must be described in the same manner as the microsystem to be measured. But this assumption is not as natural as it might seem. An object used for measuring has to be prepared and used in a special manner so if there is to be a qualitative difference between the system to be measured and the measuring apparatus. If one accepts the "natural" assumption then, for this reason alone, one has to reject all of physics, since, for instance, this physics assumes the existence of precise measurements of length and thus of rigid measuring rods, while quantum and relativity theories say that such objects do not exist. Our point is that it is no absurdity to describe an object by special rules and laws once it is used (not imagined but used!) as a measuring apparatus. In measuring one must construct measuring apparata which, in principle, enable one to perform arbitrarily exact measurements of those observables which are predicted to be exactly measurable by the theory under consideration. Bohr realised that, in quantum mechanics, the measuring apparatus has to be constructed according the rules of classical mechanics. (This postulate states more than only to be of macroscopic nature.) It is true that, from quantum theoretical perspective, which views an object as a system whose complete dynamics has to be described, the classical description is only approximate. However, in its role as measuring device, the object has to be constructed and must performed according to other rules, namely to those governed by classical physics. As to quantum mechanics, this means that in cases where the quantum structure of the measuring apparatus makes it impossible to describe it classically, the object does not work as suitable measuring apparatus. One must then tinker with the apparatus in order to be able to describe it again classically. In particular, to perform an optimal measurement one has to get up the apparatus in such a way that it behaves as if it consisted of infinitely many particles. Otherwise, any physical instrument is made of a finite number of particles. For that reason alone, the above considerations show that there are limitations to the accuracy of measurement. But, since the number of particles can be arbitrarily increased, one can improve the accuracy to the extent where it behaves practically as if it were made of an infinitely many, and the presumably eliminated cut and collapse of the wave function again occur. Optimal measurements do not eliminate the cut they are even based on it, and if one wants to

[41] See Ref. 8, p. 52.

investigate the relation of theory and experiment, then, of course, one has to consider optimal measurement.

The cut is an emergency solution producing all the quantum-mechanical difficulties, like the reduction of the state vector; nevertheless the cut characterises the essential novelty brought into physics by quantum mechanics. The cut characterises the essential novelty but not the difference between quantum mechanics and the everyday world often falsely identified with classical mechanics.

To avoid misunderstandings, it should be stressed the following: The insight one gains by quantum mechanics is not that the spaces of measurement and of dynamical representation like the configuration space must be disintegrated in such a manner that a "distortion-free projection" of one on the other is permanently excluded. It is, in principle, possible that the future development of physics expands the space of experience so that the cut between the measurement and representation spaces is eliminated or defused. The insight one gains by quantum mechanics is that these two spaces are categorically different. And this insight would also remain valid in the case that a distortion-free projection were possible again.

From this point of view, one can appraise positively those results of quantum mechanics, which usually are expressed in the following thesis: Unfortunately, due to the fact that we are macroscopic beings, we are compelled to use classical-mechanical measurement apparata, and from this all the difficulties result. For it comes out that classical mechanics proving to be a means has a more general meaning as one assumed before. Classical mechanics is certainly not general dynamics, but it is a central part of measurement theory. This can be admitted as a positive result. Of course, one only appreciates it as such if one has comprehended the necessity of a means of cognition for physical cognitive process (and, of course, for each other cognitive processes). Only this comprehension can overcome the shock on quantum mechanics. The discussion between Einstein and Bohr contributed essential points to clarify this epistemological problem and, thereby, to unify the two "profound truths", formulated as opposites, in a single truth.

To repeat: When Heisenberg and subsequent authors talk about measurement apparata, the features of these apparata are often described as classical-illustrative objects of the space-time world surrounding us. It is not explicitly said that this means: the measurement apparata must be constructed and described according to the rules of classical physics. Thus it is not cleared up that, in measuring one works not with objects one finds in our environment, but with apparata that one has to construct by ideation according to those rules. This situation implies that one states that the influence of the measurement apparatus on the system whose quantities are to be measured must be treated differently from the mutual influence of the parts of a system, but this influence is not specified as the coupling between a quantum or atomic system and a especially prepared measurement apparatus. Therefore, in this context, the space-time concept of classical physics is often denoted as a concept of everyday experience, which renders it possible to describe the experiments and their results as one does things and events in everyday life.

Only by avoiding the identification of classical-mechanical with everyday experience can the measurement means be grasped as a necessary third instance, mediating the observer and the physical object, such that the limitations of the movability of Heisenberg's cut can be determined.

Maybe, there will be a change of physics, but there is no epistemological reason to demand such change. In contrast to Penrose's,[42] our view does not force one to be a believer in some form of many-world viewpoint, it rather means to accept that we can never speak of reality in itself. But taking the means of cognition into account does not contradict a realistic view, only an ontologistic one.

4. Final Remark

Investigating the epistemological structure of physics in the period since its foundation by Newton, one finds a dualism of space-time and matter as a basic feature of physics. This dualism must be given a new definition by each physical theory. Classical mechanics determined this dualism for physics in general in so far as it occupies a special position in physics, a position that makes it necessary to refer again and again to the basic concepts developed by it. But this does not mean that the border between the two sides of the dualism could not be shifted by other physical theories.[43]

For example, the general relativity theory moved gravitational interaction from the side of dynamics to that of space-time by geometrising gravitation. Thus the dualism was redefined: the gravitational ether – a term introduced by Einstein and Weyl – replaced the class of inertial systems forming the absolute space; the gravitational ether unifies gravitation and inertia to form a new background on which all non-gravitational interactions are described and measured. The former rigid Newtonian space-time background is changed in a what Einstein called reference mollusc (*Bezugsmolluske*). Otherwise, because rigid measurement rods are necessary for measurement, the theory of general relativity is also forced to refer to a *quasi*-Newtonian situation if it is to establish a connection to the experiment, if it is to establish the possibility of physical interpretation.[44] As long as only relatively weak gravitational fields are concerned, this is easily done. However, in the case of strong gravitational fields, the modelling of the reference mollusc by means of *quasi*-rigid Newtonian space-time is not possible, not even approximately. The resulting difficulties linked with the revealing of the physical content of general relativity theory are one of the most important topics in discussions dealing with modern gravitational research today. What matters is establishing a connection with classical mechanics for measurement-theoretical reasons and, at the same time, elaborating a theory which would make it possible to grasp physical situations differing widely from those described by classical mechanics.

Since the geometrisation of gravitation also means a physicalisation of geometry, in that space-time structures were identified with the physical field of gravitation, not all space-time determinations are presupposed now but are partly dynamically determined.

---

[42] See Ref. 28, p. 238.

[43] For more details, see Refs. 51, 55, and the literature cited within.

[44] *Quasi*-Newtonian means that all correlations of the Newtonian theory are valid in the first approximation, and the effect of the Einstein theory can be presented as perturbations of this basic situation.

Thus the *a priori* measurement-theoretical part of general relativity theory was reduced in comparison to that in classical mechanics.

The need to determine the dualism of space-time and dynamics in a new manner is, as shown above, especially evident in quantum mechanics. Let us repeat: The basic dynamic law of that theory, the Schrödinger equation, can only be formulated in a many-dimensional configuration space which is fundamentally different from the space determined measurement-theoretically. Heisenberg called this dualism of space-time and dynamics even a cut. In order to link with experience, that is, to establish a connection between dynamics and measurement theory, axioms relating to the measurement process are added to the Schrödinger equation. They indicate how to arrive at the three- or four-dimensional space-time of measurement via the projection of the multi-dimensional configuration space.

Here we are dealt with the case that space in which basic physical equations are formulated does not coincide with the three-dimensional space of experience where the measurement apparata are at work. In classical mechanics, the special case arises that dynamics is related to this measurement space (the absolute Newtonian space). As long as one had only classical physical theories, the opinion prevailed that both kinds of space were identical. One knew, of course, that mathematically abstract spaces can be introduced that allow a more elegant formulation of dynamics under specific conditions. However, it was always possible to formulate dynamics relating to the measurement space, too. This situation is different in quantum mechanics.

The dualism of space-time and dynamics leads "from the determination of Schrödinger's wave equation in the configuration space to probabilistics in the physical space".[45] The probabilistic aspects of quantum mechanics result from a linkage of the Schrödinger equation (or its relativistic generalisation like Klein-Gordon equation, Dirac equation or Fock equation) to measurement. Thus, it is clearly a misinterpretation to say that so-called statistical causality brings a subjective element into the description of phenomena. This epistemological error,[46] which the founders of the quantum mechanics did not make, results from an insufficient analysis of the conditions under which phenomena in physics are studied.

These considerations invalidate arguments saying that quantum mechanics disproves the concept of reality. On these arguments, the concept of reality is to have been falsified because quantum mechanics proves the impossibility describing nature causally. Such argumentations start of the difference between quantum mechanics and classical mechanics and conclude that, if quantum mechanics is equivalent to modern physics, the concept of a causality finds no place in modern description of nature, and thus that the strict lawfulness is disproved. This approach, however, is an absolutization of classical mechanics, which designates only its laws as adequate to a complete, lawful description of nature. But this is not the standpoint of physics – it is that of mechanicism. On the other hand, it underestimates the importance of classical mechanics by ignoring the necessary linkage of quantum mechanics, expressed through the Bohr principle of complemen-

45 "... von der Determiniertheit der Schrödingerschen Wellengleichung im Konfigurationsraum auf die Probabilistik im physikalischen Raum" [30].
46 See, for instance, Ref. 17, p. 593.

tary and of correspondence (Ehrenfest theorems), to classical mechanics; but only this linkage makes the physical interpretation of quantum mechanics possible. This underestimation is, as Rosenfeld put it, an epistemological error!

One might well say that: Einstein's dissatisfication with *quantum* mechanics did not result from his considering this physical theory as indeterministic (which he did not do). He rejected the attempts of Bohm and de Broglie as being too simplistic attempts at describing the quantum mechanics deterministically, and he denied having used the question "Is the theory strictly deterministic?" as a criterion for an admissible theory [14, 26][47]. As he said, his concern was the so-called "real description". Einstein regarded the quantum mechanics as an incomplete and indirect description of reality, which would be replaced by a complete and direct one later [12]. He hoped that the development of physics would result in a theory in which measurement and representation spaces would coincide. Einstein's effort aiming at achieving a real description did not arise from a naive philosophical conception, for example, a mechanistic one, but from a profound physical program. His aim was to link modern physics to classical physics without any sharp cuts by reformulating dynamics with a immediately reference to the experience space.[48]

All discussions about quantum mechanics concentrate on solving the problem of the cut between systems to be studied and the measurement apparatus, a problem which first occurred *explicitly* in quantum mechanics. Answers must be given to questions such as, for example: Why does the cut exist? Where should it to be made? Must it occur in the sharpness given by quantum mechanics? As this cut is due to different spaces (dynamic space and experience space), the attempt to unsharpen it results in a revision of the respective concepts of space. By founding the theories of special and general relativity, Einstein just had made an extremely successful revision of space-time, and thus of the space of physical experience, and he hoped to achieved further success when he continued his works in this direction. To this end, he developed his program of a *unified* geometric field theory.

We can only judge the extent to which a modification of the experience space is actually possible, by means of real measurement possibilities. What matters is not the space of everyday experience, but the space of scientific (measuring) experience of physics. Thus it is not enough to designate the abstract mathematical configuration space as the physical experience space, but this new space must be linked with an expansion of the presupposed measurement basis of physics.

In sum: In discussing typical features of classical mechanics, the theory of relativity, and quantum mechanics one sees that measurement-theoretical requirements find their expression in the appearance of different dualisms. On these bases, physics was founded as discipline that is able to measure and to calculate motion. In consequence, the dualisms are *characteristics* of physics. They can have different forms in the respective theories but they can never be avoided. In particular, one can expect that in the future the basic dualism of geometry and physical dynamics or Heisenberg's cut, which is typical for present quantum mechanics, can be weakened. But the necessity of the dualism im-

---

47 In his letter to Born, Pauli brings the dispute between Born and Einstein to a solution.

48 Compared with the discussion between Einstein and Bohr, Boltzmann had used quite analogous arguments against the mechanics elaborated by Heinrich Hertz.

poses limitations on physical concepts like geometrodynamics, cosmology, and quantum theory without any cut.

5. References

1. A. Aspect, P. Grangier, and G. Roger, "Experimental Realization of Einstein-Podolsky-Rosen-Bohm Gedankenexperiment: a New Violation of Bell's Inequalities", *Phys. Rev. Lett.* **49** (1982) 91.
2. A. Aspect and P. Grangier, "Experiments on Einstein-Podolsky-Rosen Type Correlations with Pairs of Visible Photons", in *Quantum Concepts in Space and Time*, ed. R. Penrose and C. J. Isham (Oxford U.P., Oxford, 1986).
3. J. D. Barrow and F. J. Tipler, *The Anthropic Cosmological Principle* (Oxford U.P., Oxford, 1988).
4. A. Barut, "Schrödinger's Interpretation of π as a Continuous Charge Distribution", *Annalen der Physik (Leipzig)* **45** (1988) 31.
5. N. Bohr, "Discussion with Einstein on Epistemological Problems in Atomic Physics", in *Albert Einstein: Philosopher-Scientist*, ed. P.A. Schilpp (Open Court and Cambridge U. P., London–La Salle, 1970) pp. 199-241.
6. N. Bohr, "Über Erkenntnisfragen der Quantenphysik", in *Max-Planck-Festschrift 1958* (Dt. Verlag Wissenschaften, Berlin, 1959) pp. 169-175.
7. N. Bohr and L. Rosenfeld, "Zur Frage der Meßbarkeit der elektromagnetischen Feldgrößen", *Kgl. Danske Vidensk. Selskab. Mat-fyz. Meddelelser* **12** (1933), No. 8; English translation in *Niels Bohr, Collected Works*, Vol. 7, ed. J. Kalckar (North-Holland, Amsterdam et al., 1996) pp. 123-166.
8. M. Cini, "Quantum Theory of Measurement without Wave Packet Collapse", *Nuovo Cimento* **73B** (1983) 27.
9. A. Einstein, "Maxwell's Influence on the Development of the Conception of Physical Reality", in *James Clerk Maxwell. A Commemoration Volume. 1831-1931* (Cambridge U. P., Cambridge, 1931) pp. 66-73.
10. A. Einstein, "Physik und Realität", in *Journ. Franklin Inst.* **221** (1936) 313.
11. A. Einstein, "Quantenmechanik und Wirklichkeit", in *Dialectica* **2** (1948) 320.
12. A. Einstein "Letter to M. Born, April 5, 1948", in *Albert Einstein – Hedwig und Max Born. Briefwechsel 1916-1955, commented by M. Born* (Ullstein, Frankfurt a.M. et al., 1969) pp. 228-234.
13. A. Einstein, "Remarks to the Essays Appearing in this Collective Volume", in *Albert Einstein: Philosopher-Scientist*, ed. P.A. Schilpp (Open Court and Cambridge U. P., London–La Salle, 1970).
14. A. Einstein "Letter to M. Born, May 12, 1952", in *Albert Einstein – Hedwig und Max Born. Briefwechsel 1916-1955, commented by M. Born* (Ullstein, Frankfurt a.M. et al., 1969) pp. 257-258.
15. A. Einstein, B. Podolsky, and N. Rosen, "Can Quantum-Mechanical Description of Physical Reality Be Considered Complete?", *Phys. Rev.* **47** (1935) 777.
16. H. Everett, "'Relative State' Formulation of Quantum Mechanics", *Rev. Mod. Phys.*

29 (1957) 454.
17. C. George, I. Prigogine, and L. Rosenfeld, "The Macroscopic Level of Quantum Mechanics", in *Selected Papers of Léon Rosenfeld*, ed. R. S. Cohen and J. Stachel (Reidel, Dordrecht et al., 1979) p. 593.
18. W. Heisenberg, *The Physical Principles of the Quantum Theory* (Univ. of Chicago Press, Chicago, 1930).
19. W. Heisenberg, "Ist eine deterministische Ergänzung der Quantenmechanik möglich?", in *Wolfgang Pauli, Scientific Correspondence with Bohr, Einstein, Heisenberg, and Others*, Vol. II, ed. K. v. Meyenn (Springer, Berlin et al., 1985) pp. 409-418.
20. W. Heisenberg, "Prinzipielle Fragen der modernen Physik", in *Wandlungen in den Grundlagen der Naturwissenschaft* (Hirzel, Leipzig, 1942) pp. 46-47.
21. W. Heisenberg, *Physics and Beyond. Encounters and Conversations* (Harper and Row, NewYork et al., 1971).
22. I. Kant, *Kritik der reinen Vernunft*, in *Immanuel Kant, Werke*, ed. W. Weischedel (Suhrkamp, Frankfurt a.M, 1968) Vol. III-IV.
23. H. Margenau, "Einstein's Conception of Reality", in *Albert Einstein: Philosopher-Scientist*, ed. P.A. Schilpp (Open Court and Cambridge U. P., London–La Salle, 1970) pp. 243-268.
24. W. Pauli, "Letter to W. Heisenberg, June 15, 1935", in *Wolfgang Pauli, Scientific Correspondence with Bohr, Einstein, Heisenberg, and Others*, Vol. II, ed. K. v. Meyenn (Springer, Berlin et al., 1985) pp. 402-405.
25. W. Pauli, "Editorial", in *Dialectica* **2** (1948) 309.
26. W. Pauli, "Letter to M. Born, March 31, 1954", in *Albert Einstein – Hedwig und Max Born. Briefwechsel 1916-1955, commented by M. Born* (Ullstein, Frankfurt a.M. et al., 1969) pp. 293-297.
27. W. Pauli, "Albert Einstein in der Entwicklung der Physik", *Phys. Bl.* **15** (1959) 244.
28. R. Penrose, *Shadows of the Mind. A Search for the Missing Science of Consciousness* (Oxford U. P., Oxford et al., 1994).
29. M. Planck, *Der Kausalbegriff in der Physik* (Barth, Leipzig, 1958) pp. 13-14.
30. R. Rompe and H.-J. Treder, *Über Physik. Studien zu ihrer Stellung in Wissenschaft und Gesellschaft* (Akademie-Verlag, Berlin, 1970) p. 69.)
31. R. Rompe and H.-J. Treder, *Grundfragen der Physik* (Akademie-Verlag, Berlin, 1980) p. 36.
32. I. Rosenthal-Schneider, *Begegnungen mit Einstein, von Laue und Planck. Realität und wissenschaftliche Wahrheit* (Vieweg, Braunschweig-Wiesbaden, 1988) pp. 53-63.
33. E. Schrödinger, "Die gegenwärtige Situation in der Quantenmechanik", *Naturwiss.* **23** (1935) 807, 823, 844; English translation in *Proceeding of the American Phil. Soc.* **124** (1980) 323.
34. E. Schrödinger, "Letter to Pauli, July 1935", in *Wolfgang Pauli, Scientific Correspondence with Bohr, Einstein, Heisenberg, and Others*, Vol. II, ed. K. v. Meyenn (Springer, Berlin et al., 1985) pp. 406-407.
35. E. Schrödinger, "Die Besonderheit des Weltbildes der Naturwissenschaft" *Acta*

Phys. Austr. **1** (1947) 13; see also in *Erwin Schrödinger, Gesammelte Abhandlungen* ed. Austrian Academy of Science, (Verlag Öster. Akad. Wiss and Vieweg, Vienna, 1984) Vol. IV, pp. 409-453).
36. E. Schrödinger, *Science and Humanism*, in *Nature and the Greeks/ Science and Humanism* (Cambridge U. P., Cambridge, 1996).
37. E. Schrödinger, *Nature and the Greeks*, in *Nature and the Greeks/ Science and Humanism* (Cambridge U. P., Cambridge 1996).
38. E. Schrödinger, "What is a Law of Nature?", in *Science Theory and Man* (Dover, NewYork, 1957) pp. 133-147.
39. E. Schrödinger, *Mind and Matter* (Cambridge U. P., Cambridge, 1967) pp. 126-137.
40. H.-J. Treder, "Newton und die heutige Physik", in *Newton-Studien* (Akademie-Verlag, Berlin, 1978) pp. 7-17.
41. H.-J. Treder, *Große Physiker und ihre Probleme* (Akademie-Verlag, Berlin, 1983) p. 226.
42. H.-J. Treder, *Ann. Phys. (Leipzig)* **45** (1988) 255.
43. H.-J. Treder and H.-H. v. Borzeszkowski, "Interference and Interaction in Schrödinger's Wave- Mechanics", *Found. Phys.* **18** (1987) 77.
44. H.-H. v. Borzeszkowski, "Kantscher Raumbegriff und Einsteins Theorie", *Dt. Zs. Philosophie* **40** (1992) 36.
45. H.-H. v. Borzeszkowski and H.-J. Treder, The Meaning of Quantum Gravity (Reidel, Dordrecht, 1988.
46. H.-H. v. Borzeszkowski and R. Wahsner, *Newton und Voltaire. Zur Begründung und Interpretation der klassischen Mechanik* (Akademie-Verlag, Berlin, 1980).
47. H.-H. v. Borzeszkowski and R. Wahsner, "Zur Beziehung von experimenteller Methode und Raumbegriff", *Dt. Zs. Philosophie* **28** (1980) 685.
48. H.-H. v. Borzeszkowski and R. Wahsner, "Erwin Schrödingers Subjekt- und Realitätsbegriff", *Dt. Zs. Philosophie* **35** (1987) 1109.
49. H.-H. v. Borzeszkowski and R. Wahsner, "Quantum Mechanics and the Physical Reality Concept", *Found. Phys.* **18** (1988) 669.
50. H.-H. v. Borzeszkowski and R. Wahsner, "Heisenberg's Cut and the Limitations of its Movability", *Ann. Phys. (Leipzig)* **45** (1988) 522.
51. H.-H. v. Borzeszkowski and R. Wahsner, *Physikalischer Dualismus und dialekti scher Widerspruch. Studien zum physikalischen Bewegungsbegriff* (Wiss. Buchgesellschaft, Darmstadt, 1989).
52. *Messung als Begründung oder Vermittlung? Ein Briefwechsel mit Paul Lorenzen über Protophysik und ein paar andere Dinge*, ed. H.-H. v. Borzeszkowski and R. Wahsner (Academia, Sankt Augustin, 1995).
53. H.-H. v. Borzeszkowski and R. Wahsner, "Die Natur technisch denken? Zur Synthese von τέχνη und φύσις in der Newtonschen Mechanik oder das Verhältnis von praktischer und theoretischer Mechanik in Newtons Physik", in: *A Collective Volume in Honour of Paul Lorenzen*, ed. M. Weingarten (in print); see also *Preprint 87 of the Max Planck Institute for the History of Science* (Berlin, 1998).
54. H.-H. v. Borzeszkowski and R. Wahsner, "Not Even Classical Mechanics Is Mechanistic" in *Worldviews and the Problems of Synthesis; the yellow book of*

Einstein Meets Magritte, ed. D. Aerts, J. van der Veken, and H. van Belle (Kluwer, Dordrecht–NewYork, in print); see also *Preprint 99 of the Max Planck Institute for the History of Science* (Berlin, 1998).

55. H.- H. v. Borzeszkowski and R. Wahsner, "Dualism of Physical Space-Time and Dynamics as a Consequence of the Possibility to Measure and Calculate Motion", in *Proceedings of the Conference "Physical Interpretation of Relativity Theory", 11-14th September 1998*, ed. M. C. Duffy (British Society for the Philosophy of Science, London, 1998) pp. 56-62.
56. R. Wahsner, *Mensch und Kosmos. Die copernicanische Wende* (Akademie-Verlag, Berlin, 1978) pp. 12-31.
57. R. Wahsner, *Das Aktive und das Passive. Zur erkenntnistheoretischen Begründung der Physik durch den Atomismus – dargestellt an Newton und Kant* (Akademie-Verlag, Berlin, 1981).
58. R. Wahsner, "Erkenntnistheoretischer Apriorismus und die neuzeitliche Physik", *Dt. Zs. Philosophie* **40** (1992) 24.
59. R. Wahsner, "Das notwendige Dritte", in *Proceedings of the Eighth International Kant Congress. Memphis, 1995,* Volume II, Part 1, ed. H. Robinson (Marquette U. P., Milwaukee, 1995) pp. 389-396.
60. R. Wahsner and H.-H. v. Borzeszkowski, *Die Wirklichkeit der Physik. Studien zu Idealität und Realität in einer messenden Wissenschaft* (Peter Lang, Frankfurt a.M. et al., 1992.
61. R. Wahsner and H.-H. v. Borzeszkowski, "Vorwort", "Einleitung", "Anmerkungen", "Anhang" in *Voltaire, Elemente der Philosophie Newtons/ Verteidigung des Newtonianismus/ Die Metaphysik des Neuton*, ed. R. Wahsner and H.-H. v. Borzeszkowski (de Gruyter, Berlin, 1997).
62. J. A. Wheeler, "Genesis and Observership", in *Foundational Problems in the Special Sciences*, ed. R. E. Butts and J. Hintikka (Reidel, Dordrecht, 1977) pp. 3-33.
63. E. P. Wigner, "Remarks on the Mond-Body Question", in *Symmetries and Reflections* (Indiana U. P., Bloomington, 1967) pp. 171-184.

Global Energy and Non-local Energy

Joshua N. Goldberg
Department of Physics, Syracuse University
Syracuse, NY 13244-1130

Abstract

Noethers theorem will be used to derive the differential conservation laws which result from the invariance of an action with respect to infinitesimal space-time symmetries. This will be illustrated for the electromagnetic field and then applied to the Einstein theory. We will see that the structure of the differential conservation laws will lead to a description of the global quantities in asymptoticaly Minkowskian space-times. In particular, we shall discuss gravitational radiation at null infinity. Next we will discuss the concept of a quasi-local energy which was originally introduced by Penrose. While this will be discussed, particular attention will be given to an energy definition proposed by Hawking since it is closely related to a definition for quasi-local energy which I will propose.

1. Introduction

The concept of energy has its origin in the notion of work done in moving an object or in changing the kinetic energy of an object. However, when the force acting is conservative, derivable from a potential, then the motion can be derived from Hamiltons principle and the concept of energy comes from the invariance of the action under time translations. We expect experiments, done in a context where only difference is the starting time, to lead to the same results. In a mechanical system, this symmetry gives us a quantity which we recognize as energy. For example, consider a particle whose motion is determined by the extremum of an action defined by a Lagrangian:

$$S = \int_{t_1}^{t_2} dtL(q(t), \dot{q}(t)).$$

If the entire motion is shifted by δt, we find that to first order

$$\delta S = \int_{t_1}^{t_2} dt\{\frac{d}{dt}\frac{\partial L}{\partial \dot{q}} - \frac{\partial L}{\partial q}\}\dot{q}\delta t - \frac{\partial L}{\partial \dot{q}}\dot{q}\delta t|_{t_1}^{t_2} + (L(t_2) - L(t_1))\delta t.$$

Thus, if the action is unchanged, $\delta S = 0$, and the Lagrangian equations of motion are satisfied, the Hamiltonian or energy is the same at t_1 and t_2:

$$(p\dot{q} - L|_{t_2} = (p\dot{q} - L|_{t_1},$$

$p = \frac{\partial L}{\partial \dot{q}}$. We extend this idea to Poincaré invariant field theories whose equations of motion are derivable from an invariant action. Again we identify a quantity conserved under a rigid time translation as the energy. In fact, for electromagnetism, as we shall see, the result is the time component of the Maxwell stress-energy tensor and there a similar result for Yang-Mills theories. Indeed, in Minkowski space, linear momentum and angular momentum conservation result from spatial translations and rotations, respectively. This result follows from Noethers theorem relating conserved quantities to the existence of space-time or gauge symmetries [1]. However, in general on a given curved background one may be able to define energy and momentum densities, but there is no longer an ordinary differential conservation law unless there are Killing vectors which describe the symmetries. That is, with respect to arbitrary time and space translations in a general curved space-time, in the absence of symmetries, Noethers theorem does not lead to any conservation relation unless the metric itself is dynamical.

In the Einstein theory, gravity is represented by the geometry of a dynamical curved space-time. As a result, in general there are no Killing vectors and hence no globally defined conserved quantities although, as we shall see, there are differential conservation relations. Moreover, even locally there are no well defined quantities one can relate to observable or measurable quantities we would like to call energy and momentum. However, when the space-time is asymptotically Minkowskian, we can find global quantities we can identify with energy and momentum and imagine instruments for their measurement.

In these lectures, I shall use Noethers theorem to derive the differential conservation laws which result from the invariance of an action with respect to infinitesimal symmetries including gauge transformations and space- time mappings. This will then be illustrated for the electromagnetic field. The result for Yang-Mills fields is similar but requires the introduction of the notion of a connection. As we are basically interested in the space-time symmetries in general relativity, this does not contribute to our interests. Then I will do the same for the Einstein theory. We will see that the structure of the differential conservation laws will depend on the choice of variables in the action. This will follow with a description of the global quantities in asymptoticaly Minkowskian space-times.

My principal goal, however, is to discuss the concept of a quasi-local energy - one which clearly depends on measurable (or estimable) quantities. The notion of a quasi-local energy definition was originally introduced by Penrose [2] and then by Horowitz and Schmidt [3]. The Penrose definition is based on his twistor concepts [4] while the latter is based on a global energy defined by Hawking [5]. Both of these will be discussed with particular attention to the Hawking definition since it is closely related to a definition I will propose. In order to explain the motivation for my definition, I will need to introduce you to, but not do, calculations leading to gravitational radiation at null infinity.

2. Noethers Theorem

We assume we have an action functional on a space-time manifold M endowed with a metric g, which may or may not be dynamical. The functional depends on n independent functions $\phi_A, A = 1 \cdots n$ and their first derivatives. Under a space-time diffeomorphism, the Lagrangian density defining the action is assumed to change by a total divergence. This will be true if the Lagrangian is a scalar density and also takes into account the possibility that that the Lagrangian may depend on second derivatives linearly so that they may be eliminated by grouping them in a divergence. Thus, the action may be written [6-8] (lower case Greek indices label the coordinates and have the range $0\cdots 3$)

$$S = \int_M \mathcal{L}(\phi_A, \phi_{A,\mu}) d^4x.$$

An infinitesimal symmetry mapping of the dynamical variables, which we indicate by $\bar{\delta}\phi_A$, can be a local gauge transformation or a space-time mapping of the underlying manifold. The change in the action is determined by the functions $\bar{\delta}\phi_A$, which in the case of a space-time mapping are the Lie derivatives. We find that the change in the action is given by

$$\bar{\delta}S = \int_M \Big\{\mathcal{L}^A \bar{\delta}\phi_A + \Big(\frac{\partial \mathcal{L}}{\partial \phi_{A,\mu}}\bar{\delta}\phi_A\Big)_{,\mu}\Big\} d^4x,$$

where

$$\mathcal{L}^A \equiv \frac{\partial \mathcal{L}}{\partial \phi_A} - \Big(\frac{\partial \mathcal{L}}{\partial \phi_{A,\mu}}\Big)_{,\mu} \equiv \partial^A \mathcal{L} - (\partial^{A\mu}\mathcal{L})_{\mu}.$$

If the equations of motion are to be unchanged in form, the Lagrangian density will change at most by a divergence, $Q^{\rho}{}_{,\rho}$. It follows that

$$\mathcal{L}^A \bar{\delta}\phi_A = \big[Q^{\mu} - \partial^{A\mu}\mathcal{L}\bar{\delta}\phi_A\big]_{,\mu}. \tag{2.1}$$

Thus, when the field equations are satisfied, $\mathcal{L}^A = 0$, the divergence on the right hand side vanishes.

Typically, the symmetry will depend on a certain number of descriptors λ^i(i ranges over the number of parameters) and possibly their space-time derivatives. For example,

$$\bar{\delta}\phi_A = X_{Ai}\lambda^i + X_{Ai}{}^{\mu}\lambda^i{}_{,\mu} + \cdots,$$

where the dots indicate that there may be higher derivatives. Rather than carry out the argument in full generality, we will restrict attention to space-time diffeomorphisms defined by the mapping ξ^{μ}. In this case we have for tensorial quantities

$$\bar{\delta}\phi_A = \phi_{A,\mu}\xi^{\mu} + u_{A\mu}{}^{\nu}\chi^{\mu}{}_{,\nu}.$$

The $u_{A\mu}{}^{\nu}$ are linear in the ϕ_A and otherwise only depend on constants.

After substitution of the above (2.1) becomes

$$\begin{aligned}\Big[\mathcal{L}^A\phi_{A,\mu} - (\mathcal{L}^A u_{A\mu}{}^{\nu})_{,\nu}\Big]\xi^{\mu} = \Big\{Q^{\nu} - \Big[u_{A\mu}{}^{\nu}\mathcal{L}^A + \partial^{A\nu}\mathcal{L}\phi_{A,\mu} \\ - \big(u_{A\mu}{}^{\rho}\partial^{A\nu}\mathcal{L}\big)_{,\rho}\Big]\xi^{\mu}\Big\}_{,\nu} - \Big(u_{A\mu}{}^{\nu}\partial^{A\rho}\mathcal{L}\xi^{\mu}\Big)_{,\rho\nu}\end{aligned} \tag{2.2}$$

When the field equations are satisfied, $\mathcal{L}^A = 0$, the divergence on the right hand side vanishes and we have a differential conservation law.

If the descriptors depend only on a number of constant parameters, there is a global conservation law for each parameter when the field equations have been satisfied. However, if they are arbitrary functions, then by carrying out the indicated differentiations we get a number of identities for each order of differentiation:

$$(u_{A\mu}{}^{\nu}\mathcal{L}^A)_{,\nu} - \mathcal{L}^A\phi_{A,\mu} \equiv 0;$$
$$u_{A\mu}{}^{\nu}\mathcal{L}^A + t_\mu{}^\nu - U_\mu{}^{\nu\rho}{}_{,\rho} \equiv O;$$
$$U_\mu{}^{\nu\rho} + U_\mu{}^{\rho\nu} \equiv 0.$$

In the above we have introduce the notation

$$Q^\nu := Q^\nu{}_\mu\xi^\mu$$
$$t_\mu{}^\nu := \partial^{A\nu}\mathcal{L}\phi_{A,\mu} - Q^\nu{}_\mu,$$
$$U_\mu{}^{\nu\rho} := u_{A\mu}{}^{\rho}\partial^{A\nu}\mathcal{L}.$$

The skew symmetry of $U_\mu{}^{\nu\rho}$ leads to the generalized Bianchi identities which is the first of the above identities. Likewise it says that

$$t_\mu{}^\nu{}_{,\nu} \equiv -\mathcal{L}^A\phi_{A,\mu},$$

which imposes conditions on the $Q^\nu{}_\mu$.

It is interesting that now if there are sources for the field,

$$\mathcal{L}^A = P^A,$$

the sources cannot be arbitrary, but must satisfy the conservation equation

$$\left(u_{A\mu}{}^{\nu}P^A + t_\mu{}^\nu\right)_{,\nu} = 0$$

which follows from (2.2) and the skew symmetry of $U_\mu{}^{\nu\rho}$. These are conditions on the sources and lead to the equations of motion for compact sources in general relativity [9,10].

We will see both descriptors which depend on constant parameters and on arbitrary functions in the two examples, the Maxwell field and the Einstein gravitational field, which we will now consider in detail.

i. Maxwell field.

For the Maxwell field we will take the vector potential A_μ as the field variables and assume that the space-time need not be Minkowskian although we will consider the metric to be given. The action then is ($F_{\mu\nu} = A_{\mu,\nu} - A_{\nu,\mu}$)

$$S = -\frac{1}{4}\int_M F_{\mu\nu}F_{\rho\sigma}g^{\mu\rho}g^{\nu\sigma}\sqrt{-g}d^4x.$$

For the Maxwell field there are two kinds of symmetries for us to consider. One follows from the fact that the Lagrangian is a scalar density and therefore it changes by a divergence when the space-time manifold undergoes an infinitesimal mapping. The second symmetry is that of gauge invariance. We will consider the two separately.

Under a diffeomorphism defined by the vector field ξ^ρ, we find

$$\begin{aligned}\bar{\delta} S = \int_M \Big\{ & \bar{\delta} A_\mu (\sqrt{-g}F^{\mu\nu})_{,\nu} - \left(\bar{\delta} A_\mu F^{\mu\nu}\sqrt{-g}\right)_{,\nu} \\ & - \frac{1}{2}[F_{\mu\nu}F_{\rho\sigma}g^{\mu\rho}\bar{\delta} g^{\nu\sigma} - \frac{1}{4}F_{\rho\sigma}F^{\rho\sigma}g_{\mu\nu}\bar{\delta} g^{\mu\nu}]\sqrt{-g}\Big\} d^4x. \end{aligned} \tag{2.3}$$

Since $\mathcal{L}$ is a scalar density, we also have

$$\bar{\delta} S = \int_M (\mathcal{L}\xi^\rho)_{,\rho} d^4x$$

Substitute into the above $\bar{\delta} A_\mu = A_{\mu,\rho}\xi^\rho + \xi^\rho{}_{,\mu}A_\rho$ and $\bar{\delta} g^{\rho\sigma} = -2\xi^{(\rho;\sigma)}$. If ξ^ρ is not a Killing vector, so that $\xi^{(\rho;\sigma)} \neq 0$, there is no way to separate (2.3) into an expression involving the field equations plus a divergence. Therefore, we get the trivial result $(\mathcal{L}\xi^\rho)_{,\rho} = (\mathcal{L}\xi^\rho)_{,\rho}$. That is, there is not even a global conservation law in the absence of a Killing vector.

However, when ξ^ρ is a Killing vector, so that $\bar{\delta} g^{\rho\sigma} = 0$, (2.3) can be arranged so that

$$\xi^\rho F_{\mu\rho}(\sqrt{-g}F^{\mu\nu})_{,\nu} = \left[\sqrt{-g}\xi^\rho(F_{\mu\rho}F^{\mu\nu} - \frac{1}{4}\delta^\nu{}_\rho F_{\mu\sigma}F^{\mu\sigma})\right]_{,\nu}.$$

In the above, we recognize the differential conservation law, when the field equations are satisfied, to be just the divergence of the the Maxwell stress-energy tensor weighted by ξ^ρ. In Minkowski space we have Poincaré invariance with 10 Killing vectors and the corresponding conserved quantities.

Thus we see that there is no well defined conservation relation unless there is a Killing vector. That means that there is no well defined energy density in the absence of a time-like Killing vector. We shall see that the situation is both better and worse in general relativity.

For a gauge transformation, $\bar{\delta} A_\mu = \psi_{,\mu}$, $\bar{\delta} g_{\mu\nu} = 0$ (2.2) becomes

$$\bar{\delta} S = \int_M \left\{\psi_{,\mu}(\sqrt{-g}F^{\mu\nu})_{,\nu} - \left(\psi_{,\mu}F^{\mu\nu}\sqrt{-g}\right)_{,\nu}\right\} d^4x.$$

Integrate the first term by parts to get

$$\bar{\delta} S = \int_M \left\{[\psi(\sqrt{-g}F^{\mu\nu})_{,\nu} - \psi_{,\nu}(\sqrt{-g}F^{\nu\mu})]_{,\mu} - \psi(\sqrt{-g}F^{\mu\nu})_{,\mu\nu}\right\} d^4x.$$

Note that the coefficient of ψ is an identity as is the divergence of the expression in the brackets. This corresponds to the Bianchi identities in general relativity. Clearly, the divergence term is also an identity written so that the field equations are obvious. When there are sources, the total divergence leads to the conservation of charge, $j^\mu{}_{,\mu} = 0$.

The identity $(\sqrt{-g}F^{\mu\nu})_{,\mu\nu} \equiv 0$ also gives us conservation of charge as an integral relation, Gauss law. Consider a four dimensional domain,D with two space-like caps

S_1, S_2 and a time-like boundary Σ. Form the integral of $(\sqrt{-g}F^{\mu\nu})_{,\mu\nu}$ over D. Assume that there are no sources on Σ, and apply Stokes theorem. Then one finds that

$$4\pi Q = \int_S F^{\mu\nu} n_{\mu\nu} dS$$

is conserved.

ii. Gravitational field.

The action for the Einstein equations is given by the integral over the curvature scalar:

$$S = \frac{1}{8\pi\kappa}\int_M \sqrt{-g}R d^4x,$$
$$S = \frac{1}{16\pi\kappa}\int_M R^{\alpha\beta}\wedge\theta^\gamma\wedge\theta^\delta \epsilon_{\alpha\beta\gamma\delta}.$$

This has been written in terms of the metric and an orthonormal tetrad defined by the forms θ^α,

$$ds^2 = \eta_{\alpha\beta}\theta^\alpha\otimes\theta^\beta,$$

with the connection defined by (bold face indices label the tetrad)

$$d\theta^\alpha = \theta^\beta\wedge\omega^\alpha{}_\beta$$

and the Riemann tensor by

$$\frac{1}{2}R^\alpha{}_\beta = d\omega^\alpha{}_\beta + \omega^\alpha{}_\gamma\wedge\omega^\gamma{}_\beta.$$

The relation we get from (2.1) depends on the particular dynamical variables we choose. We shall consider two sets of variables. First, we will vary the metric and the metric connection. Then we will vary the tetrad θ^α and $\omega^\alpha{}_\beta$. In this way we will get two different descriptions of the differential conservation laws which will be useful in the later discussion of quasi-local energy.

With the metric and metric connection as the variables, the Lie derivative of the action, (2.1) gives us

$$\begin{aligned}\sqrt{-g}G^{\iota\kappa}\xi_{\iota;\kappa} &= \tau^\kappa_{K;\kappa}, \\ \tau^\kappa_K &= \sqrt{-g}\left[\tfrac{1}{2}(\xi^{\iota;\kappa} - \xi^{\kappa;\iota})_{;\iota} + G^{\iota\kappa}\xi_\iota\right].\end{aligned} \tag{2.4}$$

Note that this is an identity as it must be since we identified the change in a scalar density with the change which results from the variation of its variables separately. But, we see that when the field equations are satisfied, the divergence of the right hand side vanishes. This result differs from that above for the Maxwell field on a given curved background where we obtained a differential conservation law only when the diffeomorphism was a Killing vector.

τ^{k}_{K}, as defined above, is known as the Komar energy-momentum tensor. Komar first recognized that it could be obtained from a modification of a proposal by C. Moller [11,12]. He showed that the potential was related to the Einstein field equations by subtracting the superpotential

$$U^{[\iota\kappa]}_{K} \equiv \tfrac{1}{2}\sqrt{-g}(\xi^{\iota;k} - \xi^{\kappa;\iota})$$

from the Einstein tensor, yielding the above relationship.

Note that we can write

$$U^{[\iota\kappa]}_{K}{}_{,\kappa} = G^{\iota\kappa}\xi_{\kappa} - \tau^{\iota}_{K}.$$

Clearly, since the double divergence of the superpotential vanishes identically, the divergence of the right hand side also vanishes identically. This is equivalent to the Bianchi identities. When there are sources for the gravitational field,

$$\sqrt{-g}G^{\iota\kappa} = -8\pi\kappa T^{\iota\kappa},$$

we obtain

$$\left\{8\pi\kappa T^{\iota\kappa}\xi_{\kappa} + \tau^{\iota}_{K}\right\}_{,\iota} = 0. \tag{2.5}$$

We have obtained a conservation equation which has a contribution from the matter source of the gravitational field and one from the gravitational field itself. The contribution from the gravitational field clearly depends on the choice of the vector field defining the diffeomorphism.

However, the choice of the Komar superpotential is not unique. The addition and subtraction of the divergence of any skew tensor or pseudotensor density will change the superpotential and therefore t^{κ}. Clearly, this will change what is meant by a local energy density although it may not change a global quantity. In particular, Komar showed how his construction is related to the Einstein pseudotensor. The Einstein pseudotensor depends quadratically on first derivatives of the metric. Therefore, it can be made to vanish at any point of the manifold. Thus emphasizing the difficulty of defining a local energy density for the gravitational field.

We see that by including general relativity the situation is improved in that we are able to define a differntial conservation law for an arbitrary smooth diffeomorphism. However, the price is that we are no longer able to define a unique local energy density.

If we use the tetrad and its connection as the variables, the corresponding calculation is too complicated. However, we can obtain the desired result in the following way. From the Sparling two-form [4,13,14]

$$\sigma_{\alpha} = -\frac{1}{2}\omega^{\beta\gamma}\wedge\theta^{\delta}\epsilon_{\alpha\beta\gamma\delta},$$

we find

$$\begin{aligned} d\sigma_{\alpha} &= G_{\alpha}{}^{\beta}\theta^{*}_{\beta} + t_{\alpha}, \\ \theta^{*}_{\beta} &= \frac{1}{3!}\epsilon_{\beta\gamma\delta\epsilon}\theta^{\gamma}\wedge\theta^{\delta}\wedge\theta^{\epsilon}, \\ t_{\alpha} &= \frac{1}{2}(\omega_{\alpha}{}^{\beta}\wedge\omega^{\gamma\delta}\theta^{*}_{\beta\gamma\delta} - \omega^{\beta}{}_{\epsilon}\wedge\omega^{\epsilon\gamma}\theta^{*}_{\alpha\beta\gamma}). \end{aligned} \tag{2.6}$$

From this we obtain the conservation relation written as a

$$d(\xi^\alpha G_\alpha{}^\beta \theta^{\bullet}_\beta) + d(\xi^\alpha t_\alpha - \sigma_\alpha d\xi^\alpha) \equiv 0.$$

Here again we substitute the matter tensor for the Einstein tensor and obtain a conservation relation which includes a contribution from the gravitational field. This form also has ambiguity in the need for a choice of vector field. Furthermore, one can add a closed form to t_α without changing the conservation relation while changing the meaning of a local energy density. In fact, J. Frauendiener has shown that if one uses the natural basis for the tetrads instead of an orthonormal basis, σ_α can be identified with the Freud superpotential related to the Einstein pseudotensor [15,16]. But, there is something natural about the formulation with the Sparling forms and we shall make use of them later.

3. Global Energy

We have seen that conservation relations in general relativity are always of the form

$$\sqrt{-g} G_\alpha{}^\beta \xi^\alpha = U^{[\beta\gamma]}{}_{,\gamma} - t^\beta.$$

Consider a four dimensional domain D, as we did earlier, which we imagine as a time-like cylinder Σ with two space-like caps Σ_1 and Σ_2. The double divergence β and γ of $U^{[\beta\gamma]}$ vanishes identically. If we integrate this double divergence over D, it also vanishes identically. Since the integrand is a divergence, we can apply Stokes theorem and obtain an integral over the boundary of D which also vanishes identically. However, if we apply the field equations on the time-like boundary with the assumption that the matter source vanishes on there, and then apply Stokes theorem once again on the space-like caps, we obtain ($S_{1,2} = \partial\Sigma_{1,2}$)

$$\oint_{S_2} U^{[\beta\gamma]} dS_{\beta\gamma} - \oint_{S_1} U^{[\beta\gamma]} dS_{\beta\gamma} = \int_\Sigma t^\beta d\Sigma_\beta. \tag{3.1}$$

Thus, the conserved quantity defined by the vector ξ^α is determined by the gravitational field outside the matter distribution:

$$Q = \oint_S U^{[\beta\gamma]} dS_{\beta\gamma} \tag{3.2}$$

and the change in this quantity is determined by the flux through through the time-like surface. In fact, one can even include the flux of matter through the surface Σ. Note that Q need not depend only on the gravitational field. Any matter within D contributes to Q; for example, to the energy.

The remaining question, however, is what does this quantity mean. In an asymptotically Minkowskian space-time, one has the generators of the Poincare transformation as the asymptotic symmetry group with the translations again as a normal subgroup. Thus, in taking the limit of S to be at space-like infinity the energy- momentum defined by the Komar, Freud, and the Sparling superpotentials lead to the same result. If the limit is taken to null infinity, on the other hand, the Komar potential must be modified by a skew complex which depends on the null rays which define the limit to infinity. I dont want to treat this part of the discussion in detail. It has been worked on by many people and there is full agreement as far as energy and momentum are concerned although there are still questions with respect to angular momentum.

On the other hand, if one or more Killing vectors exist, the Komar energy-momentum tensor, t^{κ}_{K}, defines a meaningful local density. In this case, the Sparling three-form also leads to a meaningful local density. However, it is tetrad dependent. If a tetrad adapted to the sphere in Minkowski space is used, one gets an infinite contribution for the energy. If one uses this as a zero point to be subtracted, then one gets the correct answer for the Schwarzschild solution. This is a problem we shall face in the discussion of quasi-local energy. A similar result is well known for the Einstein pseudotensor. However, there is a subtlety here which we will now discuss in relation to the field in the vicinity of null infinity.

I need to discuss more fully the approximate solution in the vicinity of null infinity [18-21] because it gives me the motivation for the definition of quasi-local energy I wish to propose. For my purposes it is convenient to carry out the discussion using the tetrad formulation. To make contact with the Newman-Penrose formalism, I will use a null tetrad so that the metric is given by

$$\begin{aligned} ds^2 &= \theta^0 \otimes \theta^1 + \theta^1 \otimes \theta^0 - \theta^2 \otimes \theta^3 - \theta^3 \otimes \theta^2. \\ &= \eta_{\alpha\beta}\theta^\alpha \otimes \theta^\beta \end{aligned}$$

where here and in any discussion with a null tetrad $\eta_{\alpha\beta}$ has the value, $\eta_{01} = -\eta_{23} = 1$, other components being zero. Tetrad indices will be lowered and raised with $\eta_{\alpha\beta}$ and its inverse, respectively. The connection and curvature tensor are defined by

$$\begin{aligned} d\theta^\alpha &= \theta^\beta \wedge \omega^\alpha{}_\beta, \\ \frac{1}{2}R^\alpha{}_\beta &= d\omega^\alpha{}_\beta + \omega^{\alpha\epsilon} \wedge \omega_{\epsilon\beta}. \end{aligned} \tag{3.3}$$

Essentially, these defining equations for the connection and Riemann tensor together with the Bianchi identities,

$$dR^\alpha{}_\beta = R^\alpha{}_\gamma \wedge \omega^\gamma - \omega^\alpha{}_\gamma \wedge R^\gamma{}_\beta, \tag{3.3a}$$

are the tetrad form of the Newman-Penrose equations when written in terms of the self-dual and anti-self-dual components.

It is useful to introduce the self-dual and anti-self-dual connection by

$$\begin{aligned} \omega^{\pm\alpha\beta} &= \frac{1}{2}(\omega^{\alpha\beta} \mp i\omega^{*\alpha\beta}), \\ \omega^{*\alpha\beta} &= \frac{1}{2}\epsilon^{\alpha\beta\gamma\delta}\omega_{\gamma\delta} \end{aligned}$$

so that $(\omega^{\pm\alpha\beta})^* = \pm i\omega^{\pm\alpha\beta}$. These are the spin coefficients defined by Newman and Penrose (NP).

The Sparling two-form can now be written

$$\sigma_\alpha = -i\omega^+{}_{\alpha\beta} \wedge \theta^\beta. \tag{3.3}$$

The other equations remain the same because of the cyclic identity of the Riemann tensor. However, t_α can now be written

$$t_\alpha = -2i\omega^+{}_\alpha{}^\beta \wedge \omega^-{}_{\beta\gamma} \wedge \theta^\gamma.$$

Although I will not make use of the explicit form of the tetrad, I list it here for future reference:

$$\begin{aligned} \theta^0 &= Ndu & e_0 &= N^{-1}(\partial_u + U\partial_r + X^A\partial_x^A) \\ \theta^1 &= dr - Udu & e_1 &= \partial_r \\ \theta^\alpha &= \xi^{\mathbf{A}}{}_A(dx^A - X^A du) & e_{\mathbf{A}} &= \xi_{\mathbf{A}}{}^A\partial_A \end{aligned} \tag{3.4}$$

The indices $A, B = 2, 3, \xi^{\mathbf{3}}{}_A = \bar{\xi}^{\mathbf{2}}{}_A$, and $\xi^{\mathbf{A}}{}_A\xi_{\mathbf{A}}{}^B = \delta_A{}^B$. As usual bold face indices label the tetrads.

Whether one follows the approximate solution according to Newman- Unti or Bondi and Sachs, the equations always divide into three groups: the hypersurface or constraint equations, the dynamic equations, and the supplementary equations which really are the conservation equations. For the Einstein equations, these are $G_\alpha{}^0, G_{\mathbf{2}}{}^{\mathbf{3}}$, and $G_\alpha{}^1$, respectively. The Bianchi identities tell us that in the vicinity of null infinity when the constraint and dynamic equations are satisfied, only the $1/r^3$ part of $G_{\mathbf{A}}{}^1$ and then the $1/r^2$ part of $G_0{}^1$ remain to be satisfied. We shall assume that only these conservation equations remain to be satisfied and examine what results from (2.5). This is the subtlety that was mentioned earlier. If one integrates over the field equations, the singular portion drops out and only a finite contribution remains. We shall have to take this into account when we make our definitions of the global energy and later the quasi-local energy.

At null infinity the asymptotic Poincare symmetry yield six vectors $X_N^{\mathbf{A}}\delta_A{}^\mu, N = 1\cdots 6$ which define the rotational and Lorentz invariance while the remaining invariance leads to a vector X_0 which is an arbitrary function of the x^A. Among these transformations, supertranslations, the rigid space-time translations form an invariant subgroup with vectors $X_\alpha^0\delta_0{}^\mu$. For the rotations we obtain from (2.5)

$$^3\{d(X_N^{\mathbf{A}}\sigma_{\mathbf{A}}) - X_N^{\mathbf{A}}t_{\mathbf{A}} - \sigma_{\mathbf{A}} \wedge dX_N^{\mathbf{A}}\} = 0.$$

The super index 3 indicates that only the coefficient of r^0 is to be considered in this expression since $X_N^{\mathbf{A}}$ contains a factor of r. Integrate this expression over a cylindrical slice of null infinity Σ bounded by two space-like surfaces S_1 and S_2 defined by u_1 and u_2. This leads to an equation like (3.1):

$$\oint_{S_2} {}^3\{X_N^{\mathbf{A}}\sigma_{\mathbf{A}}\} - \oint_{S_1} {}^3\{X_N^{\mathbf{A}}\sigma_{\mathbf{A}}\} = \int_\Sigma {}^3\{X_N^{\mathbf{A}}t_{\mathbf{A}} - \sigma_{\mathbf{A}} \wedge dX_N^{\mathbf{A}}\}. \tag{3.5}$$

The quantities

$$Q_N(S) = \oint_S {}^3\{X_N^A \sigma_A\}$$

contain angular momentum and center of mass information. On the two-surface $S, \sigma_2 = \omega^+{}_{232}\theta^2 \wedge \theta^3$,

$$\omega^+{}_{232} = \tfrac{1}{2}(\omega_{012} + \omega_{232}),$$

which is $-\beta$ of the NP spin coefficients.

Similarly, for the translations we obtain

$$\begin{aligned} P_\alpha(S_2) - P_\alpha(S_1) &= \int_\Sigma {}^2\{X_\alpha^0 t_A - \sigma_0 dX_\alpha^0\}, \\ P_\alpha(S) &= \oint_S {}^2\{X_\alpha^0 \sigma_0\}. \end{aligned} \tag{3.6}$$

In the above, the super index 2 indicates that only the coefficient of r^0 is considered and $X_\alpha^0 = (1, \sin\theta\cos\phi, \sin\theta\sin\phi, \cos\theta)$.

On $S, \sigma_0 = -\omega^+{}_{032} = -\omega_{032}$ and is the spin coefficient μ. This spin coefficient is the convergence of the null rays defining future null infinity. By (3.6), the the $1/r^2$ coefficient of this convergence defines the energy, but

$$\mu = -\frac{1}{2r} + O(\frac{1}{r^2}).$$

In (3.6), $-1/r$ does not appear because only the $1/r^2$ part of μ contributes. The surface S not only has the null rays of null infinity incident, but also the outgoing rays from a null surface going into the interior. The convergence of those rays, the spin coefficient ρ, has the asymptotic behavior

$$\rho = -\frac{1}{r} + O(\frac{1}{r^3}).$$

So,

$$P_\alpha = \lim_{r\to\infty} \oint_S X_\alpha^0(\mu - \tfrac{1}{2}\rho)dS. \tag{3.7}$$

This is the result we wish to take over to the discussion of quasi-local mass.

Although we have here obtained the results at null infinity in a somewhat different fashion, they agree completely with the results of Bondi-Sachs and Newman-Penrose.

4. Quasi-local mass

We have seen that every smooth diffeomorphism yields a differential conservation relation. However, the conserved quantity has a local meaning only when the space-time admits a Killing vector and a global meaning when there are asymptotic symmetries. However, there have been proposals for local and quasi-local quantities, particularly energy. Perhaps the most developed is the proposal by Roger Penrose based on his twistor program [2,4] . The twistor equation,

$$\nabla^{(A}{}_{A'}\omega^{B)} = 0,$$

is an equation for a two component spinor. (Here the spinor indices $A, B \cdots$ and the primed indices take the values 0 and 1) In Minkowski space this equation has four independent solutions. With these four solutions, Penrose constructs ten symmetric spinors ω_N^{AB}. Transvecting these spinors with certain components of the Riemann tensor and integrating over a closed space-like two-surface, Penrose defines the ten components of the angular momentum twistor. These components contain the energy- momentum and angular momentum for the linearized Einstein equations. However, these solutions do not exist in a curved space-time. But, if one projects the twistor equation onto a smooth two-surface, one again finds four independent solutions and can again construct a quantity identified with the angular momentum twistor, thus defining quasi- local quantities one may call energy-momentum and angular momentum. The big problem with this construction is the difficulty of picturing its physical meaning because of the quantity of mathematical apparatus involved and the connection with the twistor program.

A related construction based on the Sparling form has been carried out by Dougan and Mason [22]. They use a spinor construction which is built from the beginning on the weighting function. So, here the problem reduces to finding an appropriate spinor weighting which allows them to construct ten components for the angular momentum twistor. They choose the weighting spinors to be holomorphic functions. Again, this choice gives good results at infinity, but the meaning of the selection process is obscure.

I want to sketch below, three different suggestions, which do not use spinors and which have a direct physical motivation, two of which have results which are closely related. The first of these has a motivation coming from both Newtonian concepts and electrostatics. The second is an interesting suggestion by Hawking [5] and elaborated by Horowitz and Schmidt [3] while the third, my proposal, is motivated by the global results we obtained using the Sparling two-form.

At the Royaumont meeting, GR2, Felix Pirani made an interesting proposal [23] which is related to the Komar superpotential. He said, consider a hypersurface orthogonal time-like congruence with unit time-like tangent vector field $u^\alpha = dx^\alpha/ds$, s the proper time. In general, the velocity vector undergoes an acceleration

$$a^\alpha = u^\beta \nabla_\beta u^\alpha.$$

This is the acceleration needed for an observer to maintain his motion along a time like curve. It cancels the gravitational acceleration produced by what Pirani called the relative gravitational field strength,

$$E^\alpha = -u^\beta \nabla_\beta u^\alpha. \tag{4.1}$$

In analogy with Gauss' law in electromagnetism, define

$$M(S) = -\frac{1}{4\pi}\int_S E^\alpha n_\alpha dS,$$

where S is a closed space-like two surface in t = constant. Substitute (4.1) and make use of $u_\alpha E^\alpha = 0$ to obtain

$$M(S) = \frac{1}{4\pi}\int_S (u^{\beta;\alpha} - u^{\alpha;\beta})u_\beta n_\alpha dS. \tag{4.2}$$

Apply Green's theorem and get

$$M(S) = \frac{1}{4\pi}\int_{\Sigma}(u^{\beta;\alpha} - u^{\alpha;\beta})_{;\alpha}u_{\beta}d\Sigma.$$

We recognize in (4.2) the Komar superpotential and the Komar energy integral in the above.

When the observer is in free fall, moving along a geodesic, the relative gravitational field strength is zero and one expects $M(S)$ to be zero. From the fact that u^{α} is hypersurface orthogonal, we have

$$u^{\alpha;\beta} - u^{\beta;\alpha} = \lambda^{\alpha}u^{\beta} - \lambda^{\beta}u^{\alpha}$$

for some vector λ^{α}. Substituting this into (4.2) we find that $M(S) = 0$ when u^{α} is geodesic. This is possible in flat space and in homogeneous and isotropic cosmological space-times where the symmetry makes the relative gravitational field strenth zero.

Hawking solves the Newman-Penrose equations in an asymptotically expanding Friedmann solution [5]. He identifies as the mass within a two surface slice of an outgoing null surface

$$M(S) = (4\pi)^{-3/2}\mathcal{A}^{\frac{1}{2}}\int_{S}[-\Psi_2 - \sigma\lambda + \Phi_{11} + \Lambda]dS, \tag{4.3}$$

$\mathcal{A} = \int dS$ while $\Phi_{11} = -\frac{1}{4}(R_{01} + R_{23})$ and $\Lambda = R/24$ are components of the Ricci tensor which vanish in the absence of matter.

From the Newman-Penrose equations, one recognizes the integrand as one half the Gaussian curvature the two-surface r =constant in an outgoing null surface with a quantity $\mu\rho$ subtracted,

$$\frac{1}{2}K - \rho\mu = [-\Psi_2 - \sigma\lambda + \Phi_{11} + \Lambda].$$

By the Gauss-Bonnet theorem, the integral of K over a topological sphere is 4π. Horowitz and Schmidt show that

$$-8\mu\rho = \pi^{\beta}{}_{\beta\alpha}\pi^{\gamma}{}_{\gamma}{}^{\alpha}$$

where $\pi_{\alpha\beta\gamma}$ is the extrinsic curvature of the two-surface in the four- space. So, the integral of the Hawking integrand is a topological invariant plus 1/8 the square of the trace of the extrinsic curvature. The topological invariant cancels the contribution from a metric sphere in Minkowski space. Thus the integrand contains only physical quantities and would be zero in the absence of gravitational field or matter. The integrand falls off as $1/r^3$, the surface element has a factor of r^2, and $\sqrt{\mathcal{A}}/4\pi$ yields an additional factor of r so that the limit r to infinity is finite. Hawking shows that this quantity decreases in the presence of outgoing gravitational radiation. For the asymtotically Minkowski space limit, it agrees with the Bondi-Sachs mass.

Horowitz and Schmidt take (4.3) as a definition of quasi-local mass contained within a small sphere in the interior. To understand the quasi-local meaning, they go to a Riemann normal coordinate system centered on a non-singular point. They solve the radial Newman-Penrose equations along the outgoing null cone from that point. Then they form (4.3) and examine the limit as the two-surface approaches the apex of the cone. In the absence of matter, the lowest order limit is

$$M(S_r) = \frac{1}{72}\oint \Psi_0^o \bar{\Psi}_0^o dS$$

and

$$\lim_{r\to 0}\frac{1}{r^5}M(S_r) = \frac{1}{90}T_{\alpha\beta\gamma\delta}t^\alpha t^\beta t^\gamma t^\delta, \tag{4.4}$$

where t^α is the time-like vector formed from the two null vectors orthogonal to the surface ($\frac{\partial}{\partial u}$ in terms of the tetrad given in (3.4)) and $T_{\alpha\beta\gamma\delta}$ is the Bel-Robinson tensor [24,25],

$$T_{\alpha\beta\gamma\delta} = C_{\alpha\mu\gamma\nu}C_\beta{}^\mu{}_\delta{}^\nu + {}^*C_{\alpha\mu\gamma\nu}{}^*C_\beta{}^\mu{}_\delta{}^\nu,$$

$${}^*C_{\alpha\beta\gamma\delta} = \frac{1}{2}\epsilon_{\alpha\beta\mu\nu}C^{\mu\nu}{}_{\gamma\delta}.$$

The Bel-Robinson tensor is totally symmetric and trace-free. It has a structure similar to that of the Maxwell stress-energy tensor. However, it has the wrong dimensions to be identified with energy. A similar anylsis by Pirani for the Einstein pseudotensor has a $1/r^4$ limit of a quadratic expression in the Weyl tensor, but it is not the Bel-Robinson tensor or any other identifiable quantity.

In the previous section, we found that the global mass or energy is related to the spin coefficient μ which defines the convergence of the incoming null rays at infinity. One obtains the result by forming the integral of $\mu - \frac{1}{2}\rho$ and then taking the two- surface of integration to the limit of null infinity. $-\frac{1}{2}r^{-1}$ is the Minkowski space value of μ and it is also the limiting value of $\frac{1}{2}\rho$ which is the convergence of the outgoing rays normal to the two-surface. Therefore, we may think of the global energy and momentum as the surface integral over $\mu - \frac{1}{2}\rho$ weighted by the translational Killing vector, X_α, and taken to the limit of null infinity:

$$P_\alpha = \frac{1}{4\pi\kappa}\lim_{r\to\infty}\int_S X_\alpha(\mu - \rho)r^2 \sin\theta d\theta d\phi. \tag{3.7}$$

$X_\alpha = (1, \sin\theta\cos\phi, \sin\theta\sin\phi, \cos\theta)$ are the tetrad components of the four translational Killing vectors. For an asymptotic solution of the Newman-Penrose equations, this integrand is

$$r^2(\mu - \tfrac{1}{2}\rho) = -\tfrac{1}{2}\Big[\Psi_2^o + \bar{\sigma}^o\dot{\sigma}^o + \bar{\eth}^2\sigma^o + cc\Big] + O(1/r).$$

($\dot{\sigma}^o = \frac{\partial\sigma^o}{\partial u}$). The energy is obtained with $X_0 = 1$, so that the terms in $\eth^2$ and $\bar{\eth}^2$ do not contribute as the closed two-surface has no boundary. The remaining integrand is exactly the same as the Hawking integrand ($\dot{\sigma}^o = \bar{\lambda}$) except for the terms in the matter tensor.

It is this idea which we wish to take over to a quasi-local quantity. Consider two neighboring points, p_1 and p_2 with a timelike separation which are close enough that a unique geodesic t can connect them. Form the outgoing cone $\mathcal{N}_1$ from p_1 and the incoming cone $\mathcal{N}_2$ onto p_2 with p_2 later than p_1. These cones intersect in a two surface S and their interior D forms an Alexandrov neighborhood. A null coordinate system is set up by the family of outgoing null cones, u =constant, from the time-like geodesic from p_1 to p_2 and assume that u is the proper time. A foliation of two-surfaces on $\mathcal{N}_1$ is formed by the intersection of the family of incoming null surfaces from t. These define the r =constant surfaces, r being an affine parameter. From a sphere in the neighborhood of p_1, the angular labels $(\theta\phi)$ are defined. The resulting tetrad has the form given in (3.4). So, the definition I propose is that the quasi-local energy be

$$4\pi\kappa E_{ql} = \oint_{\mathcal{N}_1\cap\mathcal{N}_2} (\mu - \tfrac{1}{2}\rho)dS.$$

This definition goes along with the construction of the surface, which always leads to a metric sphere in Minkowski space, and with the choice of tetrad.

Following the calculation of Horowitz and Schmidt for considerations close to p_1., we assume that normal coordinates have been set up in the vicinity of p_1 and that the tetrad and connection have their Minkowski space values in the limit of r to zero. The N-P tetrad is modified so that $e_{\mathbf{A}} = (m, \bar{m})$ is tangent to the foliation of two- surfaces. This creates a complication in that the spin coefficient π is not zero, but seems a more natural geometrical choice. The calculation requires knowing the tetrad, ρ, and μ to a higher order in r so the results are listed in the appendix. Up to divergence terms, we obtain

$$\mu - \tfrac{1}{2}\rho = \eth Y + \tfrac{1}{72}\Psi_0^o\bar{\Psi}_0^o r^3,$$

the same result as Horowitz and Schmidt for their energy integrand. The term $\eth Y$ represents a divergence on the two-surface which contributes nothing to the integral. Thus, one finds that

$$4\pi\kappa P(S) = \oint_S (\mu - \rho)dS = \frac{1}{72}r^5 \int_S \Psi_0^o\bar{\Psi}_0^o d\Omega + O(r^6).$$

Therefore, the relation to the Bel-Robinson tensor is the same for this proposal as that shown above in (4.4).

This result with respect to the Bel-Robinson tensor is interesting more for what it tells us about that tensor than for an understanding of energy. In the course of the solution of the N-P equations, one finds from one of the Bianchi identities that

$$\bar{\eth}\eth\Psi_0^o = -2\Psi_0^o.$$

This appears to be a very limiting result. Since Ψ_0^o has spin weight +2, the spin weighted spherical harmonic is restricted to $l = 2$. However, if the conformal tensor is finite and single valued at a regular point of the space-time, then its tetrad components will likewise be finite. As a result, the angular dependence will be determined by the angular dependence of the tetrad. It follows that at p_1, Ψ_0^o will be finite and have only an $l = 2$ component. Away from p_1 other components can appear, but they will behave like r^2.

The definition I have given is closer to an observational measurement than is that of Hawking-Horowitz-Schmidt (HHS). Measurements of the bending of light look for the deviation from the flat space rays. From the pattern observed from a given lensing object, one tries to model the distribution of mass (energy). It is hard to imagine how one would observe the distortion of a sphere around a source.

What is offered here is a definition only of a quasi-local energy. However, the connection approaches the space limit like r at p_1. Therefore, the Killing equation has ten approximate solutions in the neighborhood of p_1. If we weight the Sparling two-form with these approximate solutions, we can obtain quasi-local linear and angular momentum. These quantities do not have as clear an identification as measurable quantities.

In conclusion I should point out that on a two-surface constructed as above and in the limit we have been considering, Penrose's twistor formulation will also lead to approximate solutions of the Killing equation and therefore to a similar result to what I have given.

References:

[1.] E. Noether, "Invariante Variationsprobleme", Nachr. d. Koenig. Gesell. d. Wiss. zu Goettingen, Math.-Phys. Klasse, 235-257 (1918).
[2.] R. Penrose, Proc. R. Soc. London **A381**, 53 (1982).
[3.] G. Horowitz and B. Schmidt, Proc. R. Soc. London **A381**, 215 (1982).
[4.] R. Penrose and W. Rindler, *Spinors and Space-Time*(Cambridge University Press, Cambridge, 1986).
[5.] S. Hawking, J. Math. Phys. **9**, 588 (1968).
[6.] P.G. Bergmann, Phys. Rev. **75**, 680 (1949).
[7.] J.N. Goldberg, Phys. Rev. **111**, 315 (1958).
[8.] P.G. Bergmann, Phys. Rev. **112**, 287 (1958).
[9.] A. Einstein, L. Infeld, and B. Hoffmann, Ann. of Math. **39**, 66 (1938).
[10.] J.N. Goldberg, "Equations of Motion in Gravitation", in *Gravitation: An Introduction to Current Research*, ed. L. Witten (John Wiley, New York, 1962).
[11.] A. Komar, Phys. Rev. **113**, 934 (1959).
[12.] C. Moller, Ann. Phys. **4**, 347 (1958).
[13.] G. Sparling, unpublished (1984).
[14.] J. Madore and M. Dubois-Violette, Commun. Math. Phys. **108**, 213 (1987).
[15.] W. Thirring, "Gauge Theories of Gravitation" in *Facts and Prospects of Gauge Theories*, ed. P. Urban (Springer, Vienna, 1978).
[16.] J. Frauendiener, Class. Quantum Grav. **6**, L237 (1989).
[17.] P. Freud, Ann. Math. **40**, 427 (1939).
[18.] H. Bondi, M.G.J. van der Burg, and A.W.K. Metzner, Proc. R. Soc. London A**269**, 21 (1962).
[19.] R.K. Sachs, Proc. R. Soc. London A270, 103 (1962).
[20.] E.T. Newman and R. Penrose, J. Math. Phys. **3**, 566 (1962).
[21.] E.T. Newman and T. Unti, J. Math. Phys. **3**, 891 (1962).
[22.] A.J. Dougan and L. Mason, Phys. Rev. Lett. **67**, 2119 (1991).
[23.] F.A.E. Pirani, "Gauss' Law and Gravitational Energy", in *Les Théories Relativistes de la Gravitation*, ed. A. Lichnerowicz and M-A. Tonnelat, (CNRS, Paris, 1962).

[24.] L. Bel, "La Radiation Gravitationelle" in *Les Théories Relativistes de la Gravitation*, ed. A. Lichnerowicz and M-A. Tonnelat (CNRS, Paris, 1962).
[25.] I. Robinson, unpublished.

ASPECTS OF PHYSICAL NONLOCALITY

ARTHUR KOMAR
Physics Department, Syracuse University
Syracuse, New York 13244

ABSTRACT

Nonlocality, although qualitatively intuitively clear, is not a very well defined quantitative concept for physical theories. We discuss several valid, but inequivalent, quantitative expressions of this concept, both for classical and quantum theories, and apply them in important situations. In particular we exhibit a polarization procedure for construction a large class of nontrivial, nonlocal observables for linear field theories and derive their invariant mappings. An analysis of the physics of the various inequivalent choices of boundary conditions for field quantization throws significant light on a critical question associated with the quantization program for gravitation. It also clarifies the nature of some of the physical assumptions underlying the choice of boundary conditions, as well as indicating the existence, in principle, of experimentally verifiable differences implied by such choices.

1. Introduction

One of the central unifying topics of the present Erice course is the study of the nonlocal features of classical and quantum theories. Although from a qualitative perspective we all seem to know what is intended by the term "nonlocal", upon closer quantitative analysis the concept proves to be far more elusive. We shall examine a number of different aspects of nonlocality in classical and quantum theories and examine some of their physical implications. However we do not pretend to be exhaustive.

Our primitive conceptualization of force is that of a pull or a push, that is, a direct contact force. In Aristotlean physics contact forces were presumed to be the source of all horizontal motion. In the absence of such contact forces it was assumed that motion would cease. (For example, an arrow in flight was believed to be pushed along by the turbulence behind it.) However vertical motion was thought to be a consequence of special forces that act upon the four component elements of which all matter was believed to be composed. These forces act to return the four elements to their respective proper places in the universe: air and fire belong up, water and earth belong down. Thus we see that already at that early time the effects of gravitation were understood to result from forces that were not in direct contact with the affected objects. In classic times, the electrostatic force associated with rubbed amber and the magnetic force associated with loadstones had been observed, but they seem to have been regarded as curiosities not subject to accepted natural laws.

Galileo continued to retain the Aristotleon distinction between horizontal and vertical forces. His law of inertia applied only to horizontal forces, which he assumed to lie tangential to circles, such as the great circles on the surface of the earth. It was a generation later, with the introduction of Newton's Law of Inertia, that dynamics in all three spatial directions were treated symmetrically. However that required the explicit introduction of the action-at-a-distance force of gravity in order to account for the vertical motion of terrestrial objects as well as the motion of the moon and the planets. With this radical and, at the time, controversial, new kind of force, the first clear recognition of nonlocal phenomena seems to have entered into physical theory. However, about a century later, Faraday introduced the concept of a force field, which seemed to reestablish the phenomena of gravitation, electricity and magnetism as examples of local contact forces, albeit the respective "contacts" were visually invisible. Indeed with the present advent of quantum field theories and their employment of Feynman diagrams, one can now almost "see" the local interactions taking place. It is therefore no longer appropriate to regard the employment of force fields per se as exemplars of nonlocal phenomena. In fact, for classical nonlinear field theories, such as Einstein gravitation, there is, strictly speaking, no two-body force interpretation available, thus making the question of nonlocal action-at-a-distance moot.

We shall therefore alter our perspective somewhat and confine our discussions to examining certain nonlocal structures of interest within several linear classical and quantum field theories. The sense in which the structures are decidedly nonlocal will become evident in each example. Most surprising, occasionally some structures that seem to be local in one context can appear to be nonlocal in a different context. (Our previous example of a two-body action-at-a-distance force verses a linear force field is a simple case in point.)

2. Particle and Field

Consider the well-known phenomenon of diffraction of a light beam through a grating. The wave-like feature of light becomes evident when one observes the pattern of bright and dark lines that appear on a detecting surface. It is clear that the wave, an extended structure, must pass through the many lines of the grating in order to produce the resulting pattern. If the detecting surface is not a photographic film, but rather a photoelectric detector, one can see the diffraction pattern develop due to the individual counts of specific detectors. From this perspective it would appear that light is not really a smooth wave, but rather a statistical collection of individual localized particles. The question immediately arises: if each particle of light, a photon, is localized, and thereby must move through some one specific slit of the grating, how can it be influenced by the presence and spacing of the remaining lines of the grating?

At this point we have nothing new to add to this well studied problem. We only mention it to indicate that in the context of this experiment it is clear that the Maxwell field, which describes the wave motion of light, is understood to be spread out in space, hence sensing the entire grating, while the photon is clearly understood to be a localized structure passing through only a single line in the grating and arriving at a single location on the detector.

Consider now light from a very distant galaxy. Due to the very great distance of the source from us it is clear that the light is extremely highly collimated. The light from the galaxy can also, in principle, be passed through an extremely precise color filter. Can this light be distinguished from that emitted from a very dim laser, and if so, how? The answer is provided by Glauber[1]. His solution is to detect the times of arrival of the individual photons. In the one case they will be arriving in correlated clumps while in the other case they will be arriving randomly. But which case is which?

The answer is that the laser beam photons arrive randomly while the photons from the galaxy arrive in highly correlated clumps. This answer is rather surprising if we regard the galactic emission process to be that of individual photons being emitted from specific atoms on the various stars. If we insist that the galactic emitters are undoubtedly random individual atoms in different stars, at stupendous distances from each other, it must follow that the atoms are not emitting photons, but rather wave packets. The correlated photons being observed result from the superposition of the tails of the many emitted wave packets. From this perspective it would appear that the particles should be regarded as the nonlocal entities, while the wave packets are the more nearly local entities. (From a formal quantum field theory perspective, recall that the number operator essentially has as its canonical conjugate the phase operator. Thus, since the hallmark of a laser is to have a well-defined phase, the number of particles would be expected to be undefined, and hence be randomly distributed in its beam. Conversely, the randomly emitted wave packets have no definite phase relations, thus permitting the number operator to be somewhat better defined.)

The nonlocality of the quantum particle is even more conspicuous when one considers a particle of a definite momentum, the more common situation occurring in Feynman diagrams. The particle, having a definite momentum, cannot be localized preferentially in any region of space. Despite this well-known result, it appears that in the mind's eye of even rather sophisticated physicists, the local nature of particles, particularly in the emission process, seems well ingrained.

3. Relativistic Invariance

With the advent of relativistic field theories, and the consequent finite nature of the propagation of signals, such theories have degrees of freedom that are apparently independent of their sources. Thus their interactions with particles do seem more nearly local , at least when these radiative modes are in play. In this section we wish to exhibit another form of nonlocal structure associated with such fields. It will therefore be convenient to first digress somewhat into the structure of the Poincaré group.

The Lie algebra of the group is defined by the six infinitesimal generators of the homogeneous Lorentz group, $L_{\alpha\beta}$, and the four infinitesimal generators of the translation group, P_α, satisfying the following commutator algebra (Greek indices run from 0 to 3 and Latin indices from 1 to 3; $\eta_{\alpha\beta}$ is the Minkowski metric of signature -2):

$$[L_{\alpha\beta}, L_{\gamma\delta}] = \eta_{\alpha\gamma} L_{\beta\delta} + \eta_{\beta\delta} L_{\alpha\gamma} - \eta_{\alpha\delta} L_{\beta\gamma} - \eta_{\beta\gamma} L_{\alpha\delta}$$

$$[P_\alpha, L_{\beta\gamma}] = \eta_{\alpha\gamma} P_\beta - \eta_{\alpha\beta} P_\gamma$$

$$[P_\alpha, P_\beta] = 0 \qquad (3.1)$$

Relativistic theories are most clearly stated by giving a realization of this algebra in terms of the dynamical variables of the theory. The identification of the corresponding expressions helps provide the physical interpretation of the theory[2]. For the remainder of this section it will be sufficient for our purposes to describe this process employing the Klein-Gordon scalar field theory[3]. As we desire to exhibit a canonical realization of the algebra, we shall represent the K-G field by two (canonically conjugate) scalar fields on a three dimensional manifold, $\varphi(\mathbf{x})$ and $\pi(\mathbf{x})$. (Bold face characters will be employed to denote spatial three-vectors whether or not their vector indices are suppressed.) Thus we have

$$[\varphi(\mathbf{x}), \pi(\mathbf{y})] = i^q \delta(\mathbf{x} - \mathbf{y}) \quad , \quad [\varphi(\mathbf{x}), \varphi(\mathbf{y})] = [\pi(\mathbf{x}), \pi(\mathbf{y})] = 0 \qquad (3.2)$$

where $q = 0$ for the Poisson brackets of classical field theory and $q = 1$ for the algebraic commutator brackets of quantum field theory.

A tedious but straightforward computation will confirm that the Poincaré algebra, Eq.3.1, can now be realized by the definitions

$$L_{ab} \equiv - \int (y_a\, \varphi_{,b} - y_b\, \varphi_{,a})\, \pi\, d^3y$$

$$L_{oa} \equiv \tfrac{1}{2} \int y_a \,(\pi^2 - \eta^{mn}\, \varphi_{,m}\, \varphi_{,n} + m^2 \varphi^2)\, d^3y$$

$$P_a \equiv \int \pi\, \varphi_{,a}\, d^3y$$

$$P_o \equiv \tfrac{1}{2} \int (\pi^2 - \eta^{mn}\, \varphi_{,m}\, \varphi_{,n} + m^2\, \varphi^2)\, d^3y \qquad (3.3)$$

The physical interpretation of these expressions readily follows from Noether's theorem[4]: L_{ab} generates the spatial rotations and therefore equals the angular momentum; P_o generates the time translations and therefore equals the energy; P_a generates the spatial translations and therefore equals the linear momentum; L_{oa} generates the proper Lorentz boosts and therefore defines the relativistic center of mass. The dynamics of this theory is therefore given by the commutation relations (where q=o for Hamilton's equations and q=1 for Heisenberg's equations.)

$$i^q \varphi_{,o} = [\varphi, P_o] \quad , \quad i^q \pi_{,o} = [\pi, P_o] \qquad (3.4)$$

It is easy to confirm that Eq.3.4 is strictly equivalent to the K-G equation for a scalar field in four dimensions.

We note at this point that the entire above formalism, including all of the above expressions can be taken over for the Maxwell theory by making the following slight modification in notation[3]. Replace the spatial scalar fields φ and π by spatial vector fields $\boldsymbol{\varphi}$ and $\boldsymbol{\pi}$, and, in every above expression in which they occur quadraticly, replace their product by the corresponding scalar (i.e. dot) product. (And of course set m=0.) The only significant and somewhat surprising alteration that occurs in the resulting formalism is that the Poincaré algebra will not be satisfied unless one adds the additional relation

$$K \equiv \nabla \cdot \boldsymbol{\pi} = 0 \tag{3.5}$$

Recall, however that this relation *is* one of the Maxwell equations and must be imposed in any event in order for the present theory to be equivalent to that of Maxwell. The vanishing of K is not a dynamical equation but rather a constraint on the initial Cauchy data. It generates the gauge symmetry transformation of the theory. The fact that the relativistic algebra will not close unless Eq.3.5 is satisfied seems to imply that, in canonical quantum field theories, should one abandon this constraint equation as an operator relation and only seek to retain the vanishing of its expectation value, as is sometime done[5], problems might be expected to arise with regard to the Lorentz invariance of the resulting quantum theory.

4. Nonlocality

We shall now take, essentially as definition, the contemporary view that fields in space-time do realize local interactions. Dynamical expressions such as those given in Eq.3.3 above for the constants of the motion, energy, momentum, etc., are therefore to be regarded as integrals over local, well defined densities. We now wish to examine the nature of dynamical constants of the motion that cannot be expressed as integrals over local densities, thus being, in this sense, *essentially* nonlocal in space-time.

Given a collection of arbitrary functions, $f_A(\mathbf{x} - \mathbf{y})$, where the subscript serves to distinguish the various members of the collection, we can define, for K-G theory, by means of the polarization of the energy expression, P_o, the set of double integrals

$$F_A \equiv \tfrac{1}{2} \int f_A(\mathbf{x} - \mathbf{y}) \{ \pi(\mathbf{x})\, \pi(\mathbf{y}) + \nabla_x \varphi(\mathbf{x}) \cdot \nabla_y \varphi(\mathbf{y}) + m^2\, \varphi(\mathbf{x})\, \varphi(\mathbf{y}) \} d^3x\, d^3y \tag{4.1}$$

By direct computation it is easy to determine that these essentially nonlocal quantities satisfy the relations

$$[\, F_A , P_o] = [F_A , F_B\,] = 0 \tag{4.2}$$

These functions clearly are commuting nonlocal constants of motion. However it is important to note that these constants are not in general Lorentz-invariant, since

$$[F_A, L_{o\bullet}] \neq 0 \tag{4.3}$$

Lest the reader think that these nonlocal constants are trivial or uninteresting, it is easy to confirm that for the particular choice

$$f(\mathbf{x}) = (2\pi)^3 \int (\mathbf{k}\cdot\mathbf{k} + m^2)^{-1/2} e^{i\mathbf{k}\cdot\mathbf{x}} d^3\mathbf{k} \tag{4.4}$$

the corresponding nonlocal constant, F, is, apart from possible factor ordering, the number operator! Note that, as a consequence of Eq.4.3, the numerical value of this constant is Lorentz frame dependent (as is also the value for the energy, which is obtained by selecting the special local choice: $f(\mathbf{x}) = \delta(\mathbf{x})$).

For a better insight into what has been constructed, and how further generalizations can be easily obtained, we note that while these new objects are nonlocal in space-time, they are local in momentum space. Define the Fourier transform of the scalar fields in the usual manner

$$\varphi(\mathbf{x}) \equiv 2^{-1/2} (2\pi)^{-3/2} \int (a_{\mathbf{k}} e^{i\mathbf{k}\cdot\mathbf{x}} + a^*_{\mathbf{k}} e^{-i\mathbf{k}\cdot\mathbf{x}}) (\mathbf{k}\cdot\mathbf{k} + m^2)^{-1/2} d^3\mathbf{k}$$

$$\pi(\mathbf{x}) \equiv -i\, 2^{-1/2} (2\pi)^{-3/2} \int (a_{\mathbf{k}} e^{i\mathbf{k}\cdot\mathbf{x}} - a^*_{\mathbf{k}} e^{-i\mathbf{k}\cdot\mathbf{x}}) (\mathbf{k}\cdot\mathbf{k} + m^2)^{1/2} d^3\mathbf{k} \tag{4.5}$$

where the commutation relations are given by

$$[a_{\mathbf{k}}, a^*_{\mathbf{l}}] \equiv \delta(\mathbf{k} - \mathbf{l}) \quad , \quad [a_{\mathbf{k}}, a_{\mathbf{l}}] \equiv [a^*_{\mathbf{k}}, a^*_{\mathbf{l}}] \equiv 0 \tag{4.6}$$

In terms of these transformed variables the Hamiltonian, P_o, may, when normal ordered, be expressed as

$$P_o = \int (\mathbf{k}\cdot\mathbf{k} + m^2)^{1/2} a^*_{\mathbf{k}} a_{\mathbf{k}} d^3\mathbf{k} \tag{4.7}$$

and the more general F_A may similarly be written

$$F_A = \int g_A(\mathbf{k}\cdot\mathbf{k})\, a^*_{\mathbf{k}} a_{\mathbf{k}} d^3\mathbf{k} \tag{4.8}$$

where $g_A(k)$ are, in general, arbitrary functions of a single variable, determined by its corresponding function $f_A(\mathbf{x})$. We now see very explicitly that F_A is the integral of a density local in momentum space, and that it commutes with P_o. In fact, it can now be readily recognized that if we construct tensorial functions by interpolating products of the form $\mathbf{k}_a \cdots \mathbf{k}_b$ into the integrand of Eq.4.8, the resulting expressions will again be constants of the motion, local in momentum space but, in general, highly nonlocal in space-time. (It is evident that the particular choice of interpellation, $\mathbf{k}_a$, into the integrand of Eq.4.8, yields the Fourier transformed expression for the polarization of P_a.) We note that the tensors in space-time obtained by such interpolated terms will evidently depend on higher spatial derivatives of the scalar fields. Only for very particular choices of g_A will these constants be expressible as integrals of local densities in space-time (basically when $g_A = (\mathbf{k}\cdot\mathbf{k} + m^2)^{2n+1}$,where $n \geq 0$.)

As the quantities F_A are constants of the motion they must generate canonical mappings of solutions into solutions. It is therefore of some interest to exhibit these mappings. In fact, the desire to understand the nature of such invariant mappings provided much of the motivation to study these nonlocal constants. The mappings now are easily seen to be given by

$$\delta_A\varphi(\mathbf{x}) = \int f_A(\mathbf{x} - \mathbf{y})\, \pi(\mathbf{y})\, d^3\mathbf{y}$$

$$\delta_A\pi(\mathbf{x}) = -\int f_A(\mathbf{x} - \mathbf{y})\, (-\nabla^2\varphi(\mathbf{y}) + m^2\, \varphi(\mathbf{y}))\, d^3\mathbf{y} \tag{4.9}$$

A simple check confirms that these are indeed infinitesimal invariant canonical mappings of K-G theory.

As in the previous section, we note that every expression and relation that we have exhibited for K-G theory is immediately extendable to linear field theories of higher spin merely by interpreting the field quantities φ, π, a and a* as vectorial or, more generally, tensorial objects, which when juxtaposed, are assumed to be undergoing the action of the relevant inner product relation (i.e. contraction of all corresponding tensor indices.)

5. Quantum Boundary Conditions

A very different, conspicuous and significant source of nonlocal phenomena occurs in quantum theory due to the necessary imposition of boundary conditions. Indeed, the hallmark feature of quantum theory, the occurrence of discrete spectra, depends critically on the character of the assumed boundary conditions. It is beyond the purview of this paper to go into this question in great detail, but it is instructive for our purposes to consider the common boundary condition of the so-called "quantization in a box" as applied to K-G theory. This boundary condition comes in two varieties: either the field must vanish on the walls of the box, or periodic boundary conditions are imposed on the box walls. In the latter case, to circumvent the introduction of twice as many states per linear degree of freedom available to the system, one generally compensates for this by halving the linear dimensions of the permitted wavelengths (see below.) The density of states, which enters so centrally in statistical computations, is thereby preserved. However, it is generally not recognized that the two different boundary conditions are nevertheless physically inequivalent and make, in principle, observably different predictions concerning the localizations of the quanta in the box.

Let us now illustrate for (real) K-G scalar field theory how this occurs. In order to impose the boundary condition that the fields are entirely contained within the walls of the box, we must decompose the fields somewhat differently than we did in Eq.4.5. For this purpose we shall define an orthonormal set of functions, complete for the set of all functions that vanish on the walls of a rectangular box, each of whose edges are of length L. (We shall take the range of the coordinates to be $L \geq x_i \geq 0$.)

$$U_{\mathbf{n}}(\mathbf{x}) \equiv (2/L)^{3/2} \sin(k_1 x_1) \sin(k_2 x_2) \sin(k_3 x_3) \tag{5.1}$$

where

$$\mathbf{k}_{\mathbf{n}} \equiv \mathbf{n}\pi/L \tag{5.2}$$

and the components of $\mathbf{n}$ are all positive integers. If we further define

$$\omega_{\mathbf{n}} \equiv (\mathbf{k}_{\mathbf{n}} \cdot \mathbf{k}_{\mathbf{n}} + m^2)^{1/2} \tag{5.3}$$

our new ansatz for the transforms of the scalar fields is obtained by the linear superposition of the standing waves, as follows:

$$\varphi(\mathbf{x}) \equiv 2^{-1/2} \Sigma_{\mathbf{n}} \omega_{\mathbf{n}}^{-1/2} (a_{\mathbf{n}} + a^*_{\mathbf{n}}) U_{\mathbf{n}}(\mathbf{x})$$

$$\pi(\mathbf{x}) \equiv 2^{-1/2} \Sigma_{\mathbf{n}} \omega_{\mathbf{n}}^{1/2} (-i) (a_{\mathbf{n}} - a^*_{\mathbf{n}}) U_{\mathbf{n}}(\mathbf{x}) \tag{5.4}$$

It is easily confirmed that, with the assumption of the standard commutation relations for the $a_{\mathbf{n}}$ and $a^*_{\mathbf{n}}$ analogous to Eq.4.6, the fields of Eq.5.4 satisfy a slightly modified version of the standard commutation relations, Eq.3.2. (The modification is in the definition of the δ function; it is now to be understood as the unit distribution for the set of functions that vanish on the walls of the box.)

With the choice of Hamiltonian, P_o, an expression identical to that of Eq.3.3, we see that this is once again a canonical realization of K-G theory. Expressed in terms of $a_{\mathbf{n}}$ and $a^*_{\mathbf{n}}$, P_o is found again to be the expression Eq.4.7 (with $\mathbf{k}$ given by Eq.5.2.) The resulting theory "quantized in the box" is therefore fully specified. We should note however that, although locally the field equation is identically that of Klein-Gordon theory, and therefore Lorentz invariant, since the box provides a preferred Lorentz frame, globally the quantum theory that we have constructed breaks this symmetry.

Should we instead seek to quantize this theory with periodic conditions on the walls of the box, we must enlarge the set of basis vectors $U_{\mathbf{n}}$ to include the cosine functions as well. The theory can again be constructed almost exactly as above. We need only let the components of $\mathbf{n}$ now range over the non-negative integers, and redefine $\mathbf{k}{=}\mathbf{n}2\pi/L$ (rather than as given in Eq.5.3) in order to preserve the density of states. However the probability distribution for the location of the quanta in the lowest states is strikingly different in the two cases. In the case of the insulating box, idealized by infinitely high potentials on the walls, the ground state is given uniquely by the quantum numbers $\mathbf{n} = (1, 1, 1)$. It is therefore non-degenerate, having energy $(3\pi^2/L^2 + m^2)^{1/2}$ (in natural units) and localizes that particle predominantly at the center of the box. Its first excited state is triply degenerate, of energy $(6\pi^2/L^2 + m^2)^{1/2}$. On the other hand, the periodic box has a non-degenerate ground state given by $\mathbf{n}=(0, 0, 0)$ of energy m, localized with constant probability throughout the box. The first excited state is six fold degenerate of energy $(4\pi^2/L^2 + m^2)^{1/2}$ and also permits the fields to have non-vanishing contact with the walls of the box. Furthermore, for the periodic boundary conditions, it is easily confirmed that the states that vanish on the walls of the box must also vanish in the center of the box.

Clearly the two sets of boundary conditions for the quantization of the theory correspond to radically different physical situations, a fact generally overlooked when one goes to the limit of $L \to \infty$. But even in that limit it is now evident that one should understand the underlying physics of the assumed boundary conditions; the limit of the first case, that of the vanishing boundary conditions, requires the state functions to vanish at infinity, while that of the second case, the periodic boundary conditions, requires that the state functions stay bounded at infinity. Thus the first case represents that of a closed system where the fields are produced by local sources. The second case permits radiation from unexamined sources outside the system to enter into the finite domain and interact with the local quantum system. The latter situation corresponds to admission of field modes that are not produced by local sources and are therefore required to be subject to independent field quantization.

6. Electromagnetism in a Box

The extension of our quantization procedure to fields of higher spins appears to be rather straightforward. As we have already pointed out, the scalar field variables are replaced by vector field variables and juxtaposed quadratic terms are understood to be subject to scalar product multiplication. However there remains the considerable problem of the satisfaction of the constraint equation, Eq.3.5. For the periodic boundary conditions this can be accomplished by a careful grouping of modes and the casting out of a considerable number of basis vectors. We shall not examine this case further since the methodology is rather standard in the literature (although not employing our standing wave ansatz, but rather using traveling waves.)

The much more fascinating case is the attempt to confine the Maxwell standing waves to the interior of a box. Indeed, the gauge equation prohibits that from happening, except for the very few modes that are parallel to the walls of the box! The basis vectors $U_{\mathbf{n}}(\mathbf{x})$ of Eq.5.1 are too restrictive to permit other solutions of Eq.3.5. When Planck[6] first considered the electromagnetic standing waves in a box, he implicitly required that the walls of the box be that of a grounded conductor rather than an infinite potential well. Thus, only the components of the Maxwell field parallel to the walls are required to vanish. The components parallel to the walls *must not vanish*. More specifically, with $\mathbf{k}_{\mathbf{n}}$ given as in Eq.5.2, we define the new set of (vector) bases, $\mathbf{V}_{\mathbf{n}}(\mathbf{x})$, (orthonormal for each component individually and complete for the respective class of functions as determined by their requisite boundary conditions):

$$\mathbf{V}^1_{\mathbf{n}}(\mathbf{x}) \equiv (2/L)^{3/2} \cos(k_1x_1) \sin(k_2x_2) \sin(k_3x_3)$$

$$\mathbf{V}^2_{\mathbf{n}}(\mathbf{x}) \equiv (2/L)^{3/2} \sin(k_1x_1) \cos(k_2x_2) \sin(k_3x_3)$$

$$\mathbf{V}^3_{\mathbf{n}}(\mathbf{x}) \equiv (2/L)^{3/2} \sin(k_1x_1) \sin(k_2x_2) \cos(k_3x_3) \qquad (6.1)$$

The vector field standing waves can now be defined analogous to Eq.5.4

$$\varphi(\mathbf{x}) \equiv (2)^{-1/2} \Sigma_{\mathbf{n}}\omega^{-1/2}_{\mathbf{n}} (\mathbf{a}_{\mathbf{n}} + \mathbf{a}^*_{\mathbf{n}}) \mathbf{V}_{\mathbf{n}}(\mathbf{x})$$

$$\pi(\mathbf{x}) \equiv (2)^{-1/2} \Sigma_{\mathbf{n}}\omega^{1/2}_{\mathbf{n}} (-i)(\mathbf{a}_{\mathbf{n}} - \mathbf{a}^*_{\mathbf{n}}) \mathbf{V}_{\mathbf{n}}(\mathbf{x}) \qquad (6.2)$$

(Note that there is no scalar product implied in the above definitions; the juxtaposed terms are to be understood as ordinary multiplication separately within each component.) It is easily confirmed that the constraint equation can be satisfied by the standard transversality requirement

$$\mathbf{n}\cdot\mathbf{a}_{\mathbf{n}} = \mathbf{n}\cdot\mathbf{a}^*_{\mathbf{n}} = 0 \qquad (6.3)$$

Again requiring m=0, the remainder of the discussion resumes exactly as in the K-G theory, and the quantization of the source-free Maxwell theory (not quite) confined to the grounded conducting box is readily established. It is interesting to note that, consistent with Planck's original ideas, the electromagnetic field makes essential contact with the oscillators in the walls of the box, which presumably is its source. Of particular interest to us at this point is the clearly physical nature of the Planck-like boundary conditions: it is essentially that of a grounded conductor rather than an infinite potential well that realizes the rectangular box. (The periodic boundary conditions do not really represent a rectangular box, but rather a torus; quite a different, and physically unrealistic, situation. It is not surprising that such boundary conditions produce observably different probability distributions for the resulting Maxwell quanta.)

7. Gravitation and Concluding Remarks

We have reviewed many different ways to quantify the vague but seemingly intuitive concept of nonlocality in classical and quantum physics and attempted to clarify their implied physical assumptions as well as their observational consequences. However, as we have noted at the start of this paper, we make no claim to being exhaustive. Most conspicuously, we have neglected entirely the discussion of the very difficult and relevant subject of distant quantum mechanical correlations that produce EPR-like[7], seemingly superluminal and/or nonlocal effects. Other papers in this volume are devoted to this important topic. We comment only that the connection with our discussion is established by recognizing that the statement of such problems involve entire completed experimental arrangements that must be modeled in the theory by extended boundary conditions.

In particular, in the previous section we showed that assumptions made about the boundary conditions required on the walls of a confining box observably affect the predicted probability distribution of the quanta in the box as well as the values and degrees of degeneracy of their energy levels. Furthermore, we noted that, since a confining box determines a preferred frame of reference, difficulties of interpretation might arise for relatively moving observers, particularly in the limit of infinitely large boxes. Recall that for all finite boxes the Poincaré symmetry is hidden. One should investigate whether this impinges on the structure of the vacuum state of the resulting quantum theory, particularly in the limit of infinitely large L. Totally unexamined is the question of how or whether quantum correlations can be maintained over distances large enough to require taking into account the affects of gravitational curvature.

Finally, our considerations have brought another major question into focus. When one tries to extend the quantum program to spin 2 theories (i.e. gravitation,) it is not clear what to accept as natural and appropriate boundary conditions for the required set of standing wave solutions. We know of no actual way to confine gravitation within a finite box, and formal mathematical rules seem unconvincing since their physical significance is obscure. Indeed it is doubtful that such boundary conditions exist. Should this be true, gravitational standing waves may not be possible. This seems to imply that the existence of discrete, independent gravitational field modes may not be conceptually meaningful and that quantization programs that presume their existence will not succeed.

8. References

1. R. J. Glauber, *Optica Acta* **13** (1966) 375
2. A. Komar, *Synthese* **21** (1970) 83
3. A. Komar, *Foundations of Physics* **15** (1985) 473
4. E. Noether, *Nachr. Gesl. Wiss. Göttingen* (1918) 171
5. S. N. Gupta, *Proc. Phys. Soc. (London)* **A63** (1950) 681; K. Bleuler, *Helv. Phys. Acta* **23** (1950) 567
6. M. Planck, *Theory of Heat* (Macmillan, London, 1932) p. 270
7. A. Einstein, B. Podolsky and N. Rosen, *Phys. Rev.* **47** (1935) 777

THE EPR ARGUMENT AND QUANTUM NONLOCALITY

Bidyut Kumar Datta[(1)(*)], Venzo de Sabbata[(1)] and Renuka Datta[(2)]

(1) World Laboratory, Lausanne, SWITZERLAND;
Dipartimento di Fisica dell' Universita, Via Irnerio 46, 40126 Bologna, ITALY

(2) Department of Mathematics, Bethune College, 181 Bidhan Sarani, Calcutta 700006, INDIA;
Inter University Centre for Astronomy & Astrophysics, Post Bag 4, Ganeshkhind, Pune 411007, INDIA

Abstract

We view the final remark of the EPR paper : "one would not be able to arrive at the above conclusion if two or more physical quantities are regarded as simultaneous elements of reality only when they can be simultaneously measured or predicted". On this point of view, because of the collapse of the wave function of the combined system, the EPR entangled relations (or correlations) of momenta and positions belonging to two spatially separated particles considered in the first part of the paper do not hold good simultaneously for two separate measurements. This case does not reconcile with the locality principle and hence exhibits quantum nonlocality. Moreover, it is quite compatible with uncertainty principle. It is concluded that the EPR argument does not exhibit any paradox.

Key words : *Nonlocality, locality principle, Separability principle, EPR entanglement (correlations), hypermaximal catalog, uncertainty principle, principle of complementarity.*

(*) Permanent Address : ICSC World Laboratory, Calcutta Branch, 108/8 Maniktala Main Road, Suite J/ 77, Calcutta 700054, INDIA.

1. Introduction

In 1935 A. Einstein, B. Podolsky and N. Rosen (EPR) [1] considered a hypothetical experiment for a two-particle system to demonstrate that the quantum-mechanical description of physical reality given by the wave functions is not complete. For this purpose, they introduced a necessary condition for a theory to be considered complete: "every element of the physical reality must have a counterpart in the physical theory". Next, in order to deal with such elements of physical reality they introduced sufficient condition for their existence: " if, without in any way disturbing a system, we can predict with certainty (i.e., with probability equal to unity) the value of a physical quantity, then there exists an element of physical reality corresponding to this physical quantity".

The EPR argument for two-particle system may be thought of being composed of, as viewed by Richard Healey [2], two plausible principles, *locality* and *separability*. Locality principle requires that if two systems S_1 and S_2 are spatially separated, the intrinsic properties of S_2 cannot be immediately affected by anything done to S_1 alone. On the otherhand, separability principle requires that spatially separated systems have their own states specified by their respective wave functions and the state of the combined system would be wholly determined by the states of its constituents.

2. The EPR entanglement

Einstein, Podolsky and Rosen considered the simple case of two particles such that each of the particles is characterized by one coordinate, say x_1 and x_2, respectively, and the canonically conjugate momentum p_1 and p_2. The EPR argument for a two-particle system is based on two facts. First one concerns with the fact that for a two-particle system the sum of their momenta $p_1 + p_2$ commutes with the difference of their positions $x_1 - x_2$. Then the precise knowledge of one of them does not preclude such knowledge of the other. By virtue of this fact, it is possible to define a wave function by the Dirac δ-function:

$$\psi = h\,\delta(x_1 - x_2 + x_0), \tag{1}$$

for which one may have the following correlations of the conjugate variables, momenta and positions, belonging to each particle:

$$p_1 + p_2 = 0 \Rightarrow p_2 = -p_1, \tag{2}$$

$$x_1 - x_2 = -x_0 \Rightarrow x_2 = x_1 + x_0, \tag{3}$$

x_0 being some constant. The second fact concerns with the choice of measurement; one can have a choice of measuring either the momentum or the position of any one of the pair of particles for ascertaining their simultaneous reality. Of course, this fact resides in the propagation of the correlations (2) and (3) of the variables of the two particles.

Because of the correlations (2) and (3) of the conjugate variables belonging to two spatially separated particles implied by the wave function (1), the EPR argument provides the following conclusions. Eq.(2) demonstrates that by measuring the momentum of the first particle to be p_1, one can infer a value of $p_2 = -p_1$ for the momentum of the second particle, and eq.(3) exhibits that by measuring the position of the first particle to be x_1, one can infer a value of $x_2 = x_1 + x_0$ for the position of the second particle. Thus, starting with the assumption that the wave function should yield a complete description of the physical reality, Einstein, Podolsky, and Rosen concluded that two physical quantities, momentum and position, with noncommuting operators, could have simultaneous reality, inspite of the fact that they were *not measured simultananeously*. According to their view, one would not be able to arrive at the above conclusion if two or more physical quantities are regarded as simultaneous elements of reality *only when they can be simultaneously measured or predicted*. On this point we note that Heisenberg uncertainty principle provides a quantitative measure of the reciprocal precision of quantities such as momentum and position which are *simultaneously defined*.

In 1935 Schrödinger [3], in reaction to the EPR paper, published three papers where he considered, like Einstein, Podolsky and Rosen, the simple case of a two-particle system with detailed analysis. He referred the EPR correlations (2) and (3) as entanglement (verschränkung) between two systems, which allows one to determine x_2 of the second system by measuring x_1 of the first system or to determine p_2 by measuring p_1. The EPR correlations (2) and (3) are often referred to as the EPR entanglement in current literature. Schrödinger opined that though the values of p_2 or x_2 obtained from the entangled relations (2) or (3) cannot be checked because of the destruction of entanglement after the first measurement of either p_1 or x_1, it would not be possible to disprove them either. So, he concluded that one could, in contrast to the rules of quantum mechanics, obtain a "hypermaximal" catalog due to the commutation relation:

$$[p_1 + p_2, x_1 - x_2] = 0. \tag{4}$$

An elaborate discussion of the above is given by Treder and von Borzeszkowski [4].

3. Quantum nonlocality

The usual physical interpretation of the quantum theory centres around the uncertainty principle, which can be derived [5] from the assumption that the wave function and its probability interpretation provide the most complete specification of the state of an individual system by taking account of the de Broglie relation $\mathbf{p} = h\,\mathbf{k}$, where $\mathbf{k}$ is the wave number associated with a particular Fourier component of the wave function. This derivation led us to interpret the uncertainty principle as an *inherent and irreducible limitation* on the precision with which it is correct for us even to conceive of momentum and position as simultaneously defined quantities. Moreover, with the aid of Bohr's principle of complementarity [6, 7, 8], it can be shown that the uncertainty principle can also be obtained as an *inherent and unavoidable limitation* on the precision of all possible measurements [8].

According to the Dirac – von Neumann version of the Copenhagen interpretation, which may be considered to be the standard interpretation, the wave function provides a description of the real properties of an individual system even when they are not observed and these properties would be revealed on measurement. Moreover, the description is complete. This version of the Copenhagen interpretation postulates that the observation of one of a pair of particles collapses the wave function of the combined system altering *immediately* the properties of the other, *even if it is spatially separated from the former*. Moreover, after this observation each particle would attain its own state specified by its own wave function and pair's state would be wholly determined by the states of its constituents. Thus, it is evident that the standard interpretation along with the postulate of the collapse of the pair's wave function reconciles with the separability principle, but not with the locality principle, and hence it exhibits quantum nonlocality.

Next, taking account of the above standard Copenhagen interpretation along with the postulate of the collapse of the wave function of the combined system, we view the final remark of the EPR paper (mentioned earlier): "one would not be able to arrive at the above conclusion * if two or more physical quantities are regarded as simultaneous elements of reality only when they can be simultaneously measured or predicted." On this point of view, the EPR final remark seems to provide the following conclusion. By measuring the momentum of the first particle to be p_1, one can infer a value of $p_2 = -p_1$ for the momentum of the second particle, but loses any possibility of making any inference about the position of either particle. On the other hand, by measuring the position of the first particle to be x_1, one can infer a value of $x_2 = x_1 + x_0$ for the position of the second particle, but loses any possibility of making any inference about the momentum of either particle. Because in this case the entangled relations (correlations) (2) and (3) of the conjugate variables, momenta and positions, belonging to two spatiallly separated particles cannot hold good simultaneously for two separate measurements. Therefore in this case we have the following Heisenberg relations

$$\Delta p_1 \Delta x_1 \geq \hbar , \qquad (5)$$

$$\Delta p_2 \Delta x_2 \geq \hbar , \qquad (6)$$

This case reconciles with the separability principle, but not with the locality principle, and hence it exhibits quantum nonlocality. Moreover, it is quite compatible with the uncertainty principle, which is a necessary consequence of the assumption that the wave function and its probability interpretation provide the most complete specification of the state of an individual system.

At last we like to emphasize that whenever we have Heisenberg relations like (5) and (6), or commutation relations, we cannot reconcile with locality principles and we have to do necessarily with non-locality.

This is true also when we consider the quantization of gravity which is directly connected with quantization of some geometrical objects of the space-time itself [9].

In fact, gravitation is not a force but is the curvature of space-time, so that if we

* We are forced to conclude that (see [1]) "The quantum mechanical description of physical reality given by wave functions is not cmplete."

like to quantize gravity, we have to quantize the curvature, i.e. a geometrical property of the space-time, and if we introduce also the spin (an introduction that we deem necessarily because the spin is an original property of every elementary particle like the mass) which, geometrically, is represented by torsion, we find that between curvature R and torsion Q we can write uncertainty relations like

$$\Delta Q \, \Delta R \geq L_{Pl}^{-3} , \qquad (7)$$

where L_{Pl} is the Planck lenth connected with the Planck constant $\hbar$ ($L_{Pl} = (\hbar G / c^3)^{1/2}$) or

$$[Q, R] = i(\hbar G / c^3)^{-3/2} . \qquad (8)$$

This is only symbolic, but we can show that through a quadratic Lagrangian (see [9]) torsion and curvature behave as canonical conjugate variables for which commutation relation hold. Moreover, if we introduce the geometric algebra we can write more precise uncertainty relations where the imaginary unit is substituted by a bivector which is connected with the spin.

So, we can say that with this quantization of gravity we find ourselves in a non-local description.

4. Conclusion

The usual physical interpretation of the quantum theory centres around Heisenberg uncertainty principle, which can be derived from the assumption that the wave function and its probability interpretation provide the most complete specification of the state of an individual system. But the EPR entanglement provides a hypermaximal knowledge about the conjugate variables for ascertaining their simultaneous reality. Moreover, this hypermaximal knowledge that cannot be verified experimentally contradicts the uncertainty principle. In this sense, the EPR argument which is formally an implication of the quantum entanglement of the two particles exhibits no paradox.

von Borzeszkowski and Washner [10] observe that the EPR argument does not prove the occurrence of a paradox because Einstein tacitly worked with two different conceptions of a system, namely with a quantum-mechanical one and with a space-time determination in the sense of proximity or locality principle of classical field theory. After having used the knowledge of quantum-mechanics that two particles formed a single system, Einstein continued to argue as if there existed two different systems in the sense of classical field theory or special relativity theory. This ambiguous definition of a system was criticized by Bohr and Pauli. In principle, these critiques removed the EPR paradox. It seems to be a paradox when one discusses it in terms of classical physics.

Schrödinger [11] believed that the EPR entanglement of the two particles was correct but remarked that it was a correct implication of the quantum theory. So, the EPR argument does not provide a paradox.

Acknowledgement

The authors thank H.- H. von Borzeszkowski and R. Washner for providing the preprint [10] of their joint work and making suggestions with references [3 , 4 , 11] during the discussion of this paper .

REFERENCES

[1] A . Einstein , B . Podolsky and N . Rosen , Phys . Rev . **47** , 777 (1935) .
[2] R . Healy , in " The Great Ideas Today , Encyclopaedia Britannica , Inc ." (Chicago , 1998) pp. 74 – 104 .
[3] E . Schrödinger , Naturwissenschaften **23** , 807 , 823 , 844 (1935) .
[4] H.- J. Treder and H.- H. von Borzeszkowski , Found . Phys . **18** , 77 (1988) .
[5] D. Bohm , " Quantum Theory" (Prentice Hall , Inc . , New York ,1951) p. 611 .
[6] N. Bohr , Phys . Rev . **48** , 696 (1935) .
[7] N. Bohr , " Atomic Theory and the Description of Nature" (Camb . Univ . Press , London , 1934) .
[8] D. Bohm , see ref . 5 , Chapter 5.
[9] V. de Sabbata , L. Ronchetti and B. K. Datta , " Non commuting geometry and spin fluctuations" , in these proceedings .
[10] H.- H. von Borzeszkowski and R . Washner , " Non - locality versus locality : the epistemological background of the Einstein – Bohr debate" , (preprint) to be published in these proceedings .
[11] This comment of Schrödinger was pointed out by von Borzeszkowski during the discussion of the paper .

Non-commuting geometry and Spin Fluctuations

Venzo de Sabbata[1,2], Luca Ronchetti[3]
and Bidyut Kumar Datta[1,2](*)

(1) World Laboratory, Lausanne, Switzerland
(2) Dept.of Physics, University of Bologna, Italy
(3) Dept.of Physics, University of Ferrara, Italy

Abstract

We will see if it is possible to treat some geometrical objects like the curvature and torsion as conjugate variables for which one can write uncertainty and commutation relations that indirectly have to do with the problem of nonlocality.

Starting from the possibility that curvature and torsion be conjugate variables satisfying uncertainty relations, we consider a quadratic Hamiltonian in torsion and curvature, in order to give a better basis for that conjecture. Due to closure failure (i.e. to space-time defects) we are led (through torsion) to position fluctuations while curvature gives rise to momentum fluctuations. Expressing both torsion and curvature in real space-time and noticing that a quadratic Lagrangian in curvature and torsion describes directly the quantum fluctuations at the Planck scale of space-time points (i.e.,for instance, in the early universe) as harmonic oscillators, we show that spin fluctuations are of primary importance and they include also mass fluctuations.

(*) Permanent address: ICSC World Laboratory, Calcutta Branch
108/8 Maniktala Road, Suite J/77, Calcutta 700054, India

1. Introduction

In this lecture I will like to discuss some aspects of quantum gravity in the sense that indirectly these aspects have to do with the problem of nonlocality. We think that when we face with the problem of quantize the gravity, first of all we must consider the introduction of the spin in the theory, especially when we consider the early universe where the elementary particle physics play an important role; then we must take into account that elementary particles possess both mass and spin. We know that mass is connected with curvature while spin is connected with torsion. We think that it is possible to introduce uncertainty relations or commutation relations between these two geometrical objects like the curvature and the torsion.

Firstly we will see that is possible to find these relations by some analogy with the Bohr-Sommerfeld rules and, secondly, we will show that it is possible to introduce a quadratic Lagrangian where curvature and torsion play the role of canonical conjugate variables. Moreover we will describe tensors and spinors in a geometric algebra formalism in a way that both curvature and torsion are described in a real space-time.

As regards the opportunity to find commutation relations between curvature and torsion, we have considered in some previous works [1,2,3,4,5], also in collaboration with Borzeszkowski and Treder, the possibility that curvature and torsion can be treated in analogy to the semi-classical rules of Bohr and Sommerfeld on the basis of geometrical hypothesis, enabling us to write uncertainty relations:

$$\Delta Q \, \Delta R \geq L_{Pl}^{-3} \qquad (1)$$

where L_{Pl} is the Planck length which acts as a "quantum of length".

We arrive at the relation (1) observing that one of the most important geometrical property of torsion is that a closed contour in an U_4 manifold becomes in general a non-closed contour in the flat space-time V_4. As we will see, this non-closure property that is the fact that the integral

$$l^\alpha = \oint Q_{\beta\gamma}{}^{\alpha} \, ds^{\beta\gamma} \neq 0 \qquad (2)$$

(where $ds^{\beta\gamma} = dx^\beta \wedge dx^\gamma$ is the area element enclosed by the loop) over a closed infinitesimal contour is different from zero, (l^α represents the so-called "closure failure") can be treated as defects in space-time in analogy to the geometrical description of dislocations (defects) in crystals, and this can constitute a way to go toward the quantization of gravity that means quantization of the space-time itself.

Gravity in fact, according to Einstein, is not a force but is the curvature of space-time and then we cannot quantize gravity like other forces but we must try to quantize the space-time itself.

We know also that torsion is linked to the intrinsic spin so that we can say that the defects in space-time topology should occur in multiples of the Planck length $(\hbar G/c^3)^{1/2}$ i.e.

$$\oint Q_{\beta\gamma.}{}^{\alpha} dx^\beta \wedge dx^\gamma = n \, (\hbar G/c^3)^{1/2} n^\alpha \qquad (3)$$

(n is an integer and n^α a unit vector).

Eq. (3) is analogous to the well known relation $\oint p dq = n\hbar$, i.e. the Bohr-Sommerfeld relation and would define a minimal fundamental length and a minimal time, i.e.the Planck length entering through the minimal unit of spin or action ħ and acts as a quantum of length. Moreover if we wish to connect the initial and final positions of one and the same particle we cannot avoid uncertainty associated with torsion: for a sufficiently small area element dS, uncertainty in distance between initial and final position would be $\Delta l^{\mu} = Q^{\mu} dS$. This would induce fluctuations in distance in the metric through $\Delta l = \sqrt{g_{\mu\nu} dx^{\mu} dx^{\nu}}$. So what is important are not the point themselves but the "fluctuations" in their position, i.e. the interval between them caused by a deformation of space itself through torsion. Since curvature causes relative acceleration between neighbouring test particles we have momentum uncertainty related to curvature as (see ref.[2]):

$$m a^{\mu} ds = \Delta p^{\mu} = m\, R^{\mu}{}_{\alpha\beta\gamma} \frac{dx^{\alpha}}{ds} dx^{\beta} \eta^{\gamma} = m\, c\, R^{\mu} dS \qquad (4)$$

(where η^{γ} is the separation vector between neighbouring geodesics)

So as position fluctuations are given by torsion, momentum fluctuations are due to curvature and we can interpret quantum effects (and then uncertainty principle) as consequences of space-time deformation i.e.

$$\Delta p^{\mu} \cdot \Delta x_{\mu} \geq i\, \hbar \qquad (5)$$

where

$$\Delta x^{\mu} = Q^{\mu} \Delta S \qquad (6)$$

and

$$\Delta p^{\mu} = m\, c\, R^{\mu} dS \quad . \qquad (7)$$

We see that Q (torsion) and R (curvature) play the role of

conjugate variables of the geometry (gravitational field), thus enabling us to write commutation relations between curvature and torsion (and then connected with non-locality of the space-time itself)(analogous to $[x,p] = i\hbar$) as:

$$\left[Q\ ,\ R\right] = i(\hbar G/c^3)^{-3/2} = i\ L_{Pl}^{-3} \qquad (8)$$

that can be written also as

$$\Delta Q\ \Delta R \geq L_{Pl}^{-3} \qquad (9)$$

Before to continue in considering fluctuations, we must pay our attention on the fact that when we consider the early universe we have to deal both with elementary particle physics using quantum theory and with cosmology using general relativity; but general relativity is developed in real space-time while quantum theory needs a complex manifold. In order to conciliate general relativity with quantum theory we think that one must reformulate Dirac's theory in terms of space-time geometric calculus without any complex number (see ref.[6,7]). For that reason we have considered the geometric algebra that, with the multivector concept and the interpretation of imaginary units as generator of rotations, places tensors and spinors on the same foot: both are described in a real space-time. The important fact is that the generator of rotations can be interpreted as spin (see Appendix A).

Considering this geometric algebra in the four dimensional space-time and noticing that in Dirac theory we have a totally antisymmetric spin density, we are in position to introduce the torsion tri-vector

$$Q = Q^{\alpha\beta\gamma}\gamma_\alpha\wedge\gamma_\beta\wedge\gamma_\gamma \qquad (10)$$

as element of space-time algebra, where $\{\gamma_\alpha\}$ are the base vectors for which we have

$$\gamma_\alpha \gamma_\beta = g_{\alpha\beta} + \gamma_\alpha \wedge \gamma_\beta . \tag{11}$$

Moreover, given the curvature bivector

$$\Omega^{\alpha\beta} = (1/2) R^{\alpha\beta\mu\nu} \gamma_\mu \wedge \gamma_\nu \tag{12}$$

one can form the curvature tri-vectors

$$R^\alpha = \Omega^{\alpha\beta} \wedge \gamma_\beta . \tag{13}$$

We have seen that torsion and curvature are to be considered as conjugate variables and now we are in position to write eq. (6) in a real form.

In fact we have the trivectors Q and R^α. Consider now the antisymmetric part of the geometric product

$$[Q, R^\alpha] = (1/2)(QR^\alpha - R^\alpha Q) \tag{14}$$

This type of product between two tri-vectors gives a bivector (for example it is easy to verify

$$[\gamma_0 \wedge \gamma_1 \wedge \gamma_2 \, , \, \gamma_0 \wedge \gamma_1 \wedge \gamma_3] = \gamma_2 \wedge \gamma_3 \tag{15}$$

remembering that $\gamma_0 \gamma_1 \gamma_2 = i\gamma_3$ where 'i' indicates the pseudoscalar unit. In the language of geometric algebra, the imaginary unit of complex numbers is substituted by a bivector; then we can have commutation relations of canonic type (see Appendix B) from eqs.(10,12,13,15):

$$[Q \, , \, R^a] = (1/2)(QR^\alpha - R^\alpha Q) =$$

$$= (1/2)\left[Q^{\mu\nu\beta} \gamma_\mu \wedge \gamma_\nu \wedge \gamma_\beta \, , \, R^{\alpha\tau\nu\beta} \gamma_\tau \wedge \gamma_\nu \wedge \gamma_\beta\right] =$$

$$= (1/2)(Q^{\mu\nu\beta} R^{\alpha\tau}{}_{\nu\beta} - R^{\alpha\mu}{}_{\nu\beta} Q^{\tau\nu\beta}) \gamma_\mu \wedge \gamma_\tau = \gamma_2 \wedge \gamma_1 \, L_{Pl}^{-3} \tag{16}$$

for every α, where, being the first member a bivector, also the second member is a bivector, coherently with the fact that the imaginary unit in Dirac equation is substitute by the bivector

$\gamma_1 \wedge \gamma_2$ (see, for instance [6]).

2. A quadratic Hamiltonian

In order to give a stronger basis to the consideration of torsion and curvature as canonical conjugate variables, we try to see if is possible to improve the theory introducing a quadratic Hamiltonian function of torsion and curvature. We use as conjugate variable the torsion and the curvature trivectors because the first Bianchi identity (see [6]) guarantees that it is different from zero only in the presence of torsion. Since the commutation relations between Q and R^a are defined in the geometric algebra we need to modify the usual lagrangian field theory.

We start from the results of ref.[8]: the field equation for the multivector ψ can be written in the manifestly invariant way:

$$\nabla \left(\frac{\partial \mathcal{L}}{\partial(\nabla\psi)} \right) = \frac{\partial \mathcal{L}}{\partial \psi} \tag{17}$$

where the gradient operator $\nabla = \gamma_\mu \partial^\mu$ acts as the geometric product.

The presence of ∇ or, in other hand, the peculiarity of the intrinsic geometric calculus, suggests that we can define the conjugate momentum field as

$$\Pi = \frac{\partial \mathcal{L}}{\partial(\nabla\psi)} \tag{18}$$

and the Hamiltonian

$$H = (\nabla\psi)\Pi - \mathcal{L} \tag{19}$$

This is different from the usual $H = \Pi\dot{\psi} - \mathcal{L}$ formula, but, following eq.(17) for the Lagrangian, we propose also a modified Hamiltonian that is manifestly invariant like eq.(17) so we cannot work with $\partial\psi/\partial x^0$ because it depends from the choice of the time-coordianate x^0. So $\Pi = \frac{\partial \mathcal{L}}{\partial(\nabla\psi)}$ is the natural choice; in fact eq.(17) generalizes the Lagrangian eq.$(d/dt)(\partial\mathcal{L}/\partial\dot{q}) - \partial\mathcal{L}/\partial q)=0$ simply substituting d/dt with ∇.

Notice that eq.(17) allows for vectors, tensors and spinors variables, to be handled in a single equation a considerable unification (see [8,9]).

So we have the Hamilton equations

$$\begin{cases} \nabla\psi = \partial H / \partial \Pi \\ \nabla\Pi = - \partial H / \partial \psi \end{cases} \tag{20}$$

For example we have the Maxwell equations in vacuum taking $\mathcal{L}= F^2/8\pi$ where the vector potential $A = A^{\mu}\gamma_{\mu}$ and the bivector $F = \nabla \wedge A$ are conjugate variables (remembering that the geometric product gives $F = \nabla A = \nabla\cdot A + \nabla\wedge A$ and Lorentz condition $\nabla\cdot A = 0$).

Moreover since we use the curved U_4 manifold in agreement with the minimal coupling principle, we substitute the covariant operator D for ∇ defining

$$D \equiv \nabla + \gamma_a[\omega^a, \ ] \tag{21}$$

where the $\{\gamma_a\}$ is a orthonormal frame of tetrads, (for which

$$\gamma_a\gamma_b = \eta_{ab} + \gamma_a\wedge \gamma_b), \tag{22}$$

(related with the coordinate system $\gamma_{\mu} = e^a_{\mu}\gamma_a$) (see Appendix C).

and $\omega^a = (1/2)\omega^{abc}\gamma_b\wedge\gamma_c$ is the connection bivectors.

If in the eq. (17) we put D instead of ∇, we have $D\left(\frac{\partial\mathcal{L}}{\partial(D\psi)}\right) = \frac{\partial\mathcal{L}}{\partial\psi}$.

We remember that in the tetrad basis the connection is

$$\omega^{abc} = \omega^{a[bc]} = C^{cba} - C^{bac} - C^{acb} - K^{abc} \qquad (23)$$

(where $C_{ab}{}^{c} = e^{\mu}{}_{a} e_{b}{}^{\nu} \partial_{[\mu} e_{\nu]}{}^{c}$,the Ricci connection coefficients, give the Riemannian part and K^{abc} is the contorsion tensor which is $= - Q^{abc}$ because, in this case, we are considering the totally antisymmetric torsion).

Moreover it is possible to define the covariant derivative in the direction of a vector v putting (see Appendix C)

$$D_v \equiv v \cdot D = v^a \partial_a + v^a [\omega_a \ , \quad] \qquad (24)$$

In geometric algebra the commutation product with bivector leaves unchanged the grade of multivectors [8] and then in particular we have $(\gamma_a \cdot D)\gamma_b = \omega_{ab}{}^{c}\gamma_c$ as we expect. We can separate the interior and exterior covariant derivative putting $D \equiv D_I + D_E$, where

$$D_I = \nabla \cdot + \gamma_a \cdot [\omega^a \ , \quad] \qquad (25)$$

$$D_E = \nabla\wedge + \gamma_a \wedge [\omega^a \ , \quad] \qquad (26)$$

that respectively through the inner and outer product lover and raise the grade of multivector on which operate.

The D_E operator acts like the exterior covariant derivative in the language of exterior forms, so we have the Cartan structure equations:

$$D_E \gamma_a = Q^{bc}{}_{a} \gamma_b \wedge \gamma_c \equiv Q_a \qquad (27)$$

$$D_E(\omega^{lba}\gamma_l) = (1/2) R^{cdab} \gamma_c \wedge \gamma_d \equiv \Omega^{ab} \qquad (28)$$

and the Bianchi identities

$$D_E Q^a = \Omega^{ab} \wedge \gamma_b \equiv R^a \qquad (29)$$

$$D_E \Omega^{ab} = 0 \qquad (30)$$

Now we are able to introduce the following Lagrangian in order to have the usual expression for conjugate variables in a quantum theory:

$$\mathcal{L} = (\hbar c/2)R^aR_a - (c^4/2G)Q^aQ_a \qquad (31)$$

$$\mathcal{L} = (\hbar c/2)\left(R^aR_a - Q^aQ_a/L_{Pl}^2\right) \qquad (32)$$

In agreement with the Bianchi identities, the momenta conjugate to the torsion bivectors are the trivectors:

$$\Pi^a = \partial\mathcal{L}/\partial(DQ^a) = \hbar c R^a \qquad (33)$$

(in natural unit $\hbar$=c=1 we can write Π^a or R^a without distinction) The Hamiltonian results:

$$H = DQ^a\Pi_a - \mathcal{L} = (\hbar c/2)\left(R^aR_a + Q^aQ_a/L_{Pl}^2\right) \qquad (34)$$

Then we have the Hamilton equations:

$$\begin{cases} DQ^a = \partial H/\partial R^a = R^a \\ DR^a = -\,\partial H/\partial Q^a = -\,Q^a/L_{Pl}^2 \end{cases} \qquad (35)$$

that we can write

$$DDQ^a = -\,Q^a/L_{PL}^2 \qquad (36)$$

or equivalently for R^a

$$DDR^a = -\,R^a/L_{Pl}^2 \qquad (37)$$

Separating in eqs.(35) the exterior and interior parts of different grade we have the identities

$$\begin{cases} D_E Q^a = R^a \\ D_E R^a = 0 \end{cases} \qquad (38)$$

(the second equation is a consequence of eq.(30) when the torsion is totally antisymmetric) and the field equations

$$\begin{cases} D_I Q^a = 0 \\ D_I D_E Q^a = - Q^a/L_{Pl}^2 \end{cases} \qquad (39)$$

3. Spin fluctuation

Lagrangians quadratic in curvature and torsion are nothing new (see for instance ref.[10,11,12,13]), but there is an interesting reason supporting the choice made here of the Hamiltonian density (34): it seems to describe directly the quantum fluctuations at the Planck scale of space-time points as harmonic oscillators. In fact (see eqs.(6,7)) we can take $R^a dS \approx \Delta p^a/m_{Pl}c$ and $Q^a dS \approx \Delta x^a$ as uncertainties and taking $(dS)^2 L_{Pl}^{-1} \approx dx^3$ we have $Hdx^3 \approx (\Delta p)^2/2m_{Pl} + (1/2)m_{Pl}\omega_{Pl}^2(\Delta x)^2$, that is just the Hamiltonian of a harmonic oscillator with Planck frequency $\omega_{Pl} = c/L_{Pl}$ and Planck mass $m_{Pl} = \hbar/L_{Pl}c$ (notice that we consider always energy fluctuation at the Planck scale).

The wave eq.(36) for torsion - formally a Klein-Gordon type equation for a m_{Pl} particle - could describe the propagation of the fluctuation of the background geometry due to the torsion itself: in fact this equation is not linear because the connection contains also the torsion and then acts on itself.

In this quantum physical context, propagating torsion is not in contradiction with Einstein-Cartan theory - for which it is zero in vacuum - because we have to do with vacuum polarization as source through spin fluctuations.

The quadratic Lagrangian $\mathcal{L}$ is incomplete for a gravitation theory, since it do not give the Einstein equations and we know that a quantum theory of gravity must be reduced to the classic theory when one considers gravitation far away from the Planck scale. Then we add to (31) the usual linear Lagrangian $-\overset{o}{R}/2\chi$:

$$\mathcal{L}' = \mathcal{L} - \overset{o}{R}/2\chi \tag{40}$$

where $\overset{o}{R}$ is the scalar curvature related to Riemannian part of the connection and $\chi = 8\pi G/c^4$ (i.e. $L_{Pl}^2 = \hbar c\chi/8\pi$).

Now with the tetrad field (22) as variables, we can write

$$-\overset{o}{R} = \overset{o}{R}{}^{ab}\gamma_a\gamma_b - (1/2)\gamma_a\gamma^a\overset{o}{R} \tag{41}$$

where $\overset{o}{R}{}^{ab}$ is the Ricci tensor (and $\gamma_a\gamma^a = 4$), but here it is a function of γ^a . In that way, neglecting the quadratic terms of $\mathcal{L}'$, we can have the Einstein equations using eq.(41):

$$(1/2)(\partial\overset{o}{R}/\partial\gamma_a) = \overset{o}{R}{}^{ab}\gamma_b - (1/2)\gamma^a\overset{o}{R} = \chi\ \partial\mathcal{L}_m/\partial\gamma_a \tag{42}$$

where $\mathcal{L}_m = \mathcal{L}_m(\psi,\ \gamma^a,\ Q^a)$ is the matter Lagrangian and $\partial\mathcal{L}_m/\partial\gamma_a$ defines the energy-momentum tensor $T^a = T^{ab}\gamma_b$ (see eq.(17) with D in place of ∇).

Moreover we can have the Einstein-Cartan equations simply taking γ_a and Q_a as variables and neglecting in $\mathcal{L}'$ only the term quadratic in curvature:

$$G^a = \chi\ T^a \tag{43}$$

$$Q^a = \chi\ \partial\mathcal{L}_m/\partial Q_a \tag{44}$$

where $G^a = (R^{ab} + \eta^{ab}R)\gamma_b$ is the non-symmetric Einstein tensor and $\partial\mathcal{L}_m/\partial Q_a$ defines the spin density of the matter field as torsion source.

To give eqs.(43,44) note that [14]:

$$R_{(ab)} = \overset{0}{R}_{ab} + Q_{acd}Q_b{}^{cd} = \overset{0}{R}_{ab} + Q_a \cdot Q_b \qquad (45)$$

with Q_{abc} totally antisymmetric i.e. $Q_{abc} = Q_{[abc]}$ as in our theory, and (by calculation):

$$R_{[ab]}\gamma^b = D_I Q_a \qquad (46)$$

Equation (45) implies $R = R_a{}^a = \overset{0}{R} + Q_a \cdot Q^a$ therefore $\mathcal{L}' = - R/2\chi$.

Consider now eq. (36) without executing explicity all the geometric products (inner, outer, commutator); we can achieve qualitatively important results that reveal the physical content of the equation. We write the operator D as sum of three parts: $D = \nabla + C + Q$, i.e. differential (flat) Riemannian and torsionic part respectively (see eqs.(21) and (23)). We have:

$$(\nabla + C + Q)(\nabla + C + Q)Q = - Q/L^2_{Pl} \qquad (47)$$

and, evaluating all different terms, we have equations of type:

$$\Box Q + (\nabla C)Q + C(\nabla Q) + C^2Q + CQ^2 + Q\nabla Q + Q^3 + Q/L^2_{Pl} = 0 \qquad (48)$$

(where $\Box = \nabla^2$).

At the lower grade of approximation this gives for quantum propagating torsion the Klein-Gordon equation in Minkowski space-time

$$\Box Q + Q/L^2_{Pl} = 0 \qquad (49)$$

and we expect plane wave solutions

$$Q = Q_0 \cos[(\omega_{Pl}/c)k\cdot x] \qquad (50)$$

where $\omega_{Pl}/c = 1/L_{Pl}$ and k is the wave vector ($k^2 = 1$).
Now we consider the first equation that we can have different from the Klein-Gordon type, i.e. we neglect in eq.(48) the Riemannian part and Q^3:

$$\Box Q + Q\nabla Q + Q/L_{Pl}^2 = 0 \qquad (51)$$

If we pretend that plane wave solutions are still valid, a condition is needed; moreover we note that the term with derivatives of Q is shifted in phase of $\pi/2$ respect Q, so $\nabla Q = 0$ implies

$$k\cdot x/L_{Pl} = n\pi \quad (n \text{ integer}) \qquad (52)$$

With $k^0 = 1$ (Q only time-dependent) we have the quantization of time $t = n\tau_{Pl}$ and then (if we consider the light motion) also the quantization of distances $l = nL_{Pl}$ as we have already suggested introducing space-time defects due to torsion. In fact taking $\Delta l \approx Q\,\Delta S$ in eq. (49), we have just the harmonic oscillator equation:

$$\Delta\ddot{l} + \omega_{Pl}^2 \Delta l = 0 \qquad (53)$$

that describes the fluctuations of space-time points and its confinement at the Planck lenght scale.

This is analogous to the quantum concept of Zitterbewegung as fluctuation of the order of the Compton lenght of the position of the electron (see ref.[6]). Like torsion, Zitterbewegung seems strictly related to the spin, so we can consider ω_{Pl} in eq.(53) as the analogous of the Compton frequency $\omega_c = c/r_c$ (see in particular eq.(93) in ref.[6] which reads $\Delta\theta/\Delta x^0 = RdS/Q^0dS = 1/r_c$ being r_c the Compton radius) so that we can write:

$$\omega_{Pl}/c = 1/L_{Pl} \qquad (54)$$

Note that the modified harmonic oscillator equation corresponding to eq.(51)

$$\ddot{\Delta l} + (1 + \dot{\Delta l}/c)\omega_{Pl}^2 \Delta l = 0 \qquad (55)$$

leads us to the same consideration and implies quantization of time.

One could consider eq.(52) just as the limit condition that in quantum mechanics problems gives the quantization of energy but here we work on the spacetime geometry itself. Moreover eq.(52) implies that in these points Q fluctuates between $+Q_0$ and $-Q_0$ (see eq.(50)): this fact, in agreement with Cartan theory can be interpreted as spin fluctuation between $\pm \hbar/2$ and could explain why spin can only have these two values. Then the propagation of this torsion waves gives a background of quantized space-time and spin, and is rich of consequences that are to be investigated.

At this regards we can observe that spin fluctuation are in agreement with the Pauli exclusion principle: the point is that fluctuations of the order of the Planck length render undistinguishable the extremes of fluctuation and, introducing the spin, this means that we associate the spin to the Q-wave so that we have that close extremes have opposite spin (in agreement with Pauli principle); in fact near spins (every wavelength) are ever opoosite. Then the mean value of Q over one period is always null, but this is not true for Q squared: it is typical for all wave phenomena and we know that the square of amplitude is related to the energy. These considerations, together with the geometric identity (45) suggest that the torsionic part of the curvature could be converted into the Riemannian part connected with the energy-momentum tensor through the Einstein equation:

$$R_{ab} = Q_a \cdot Q_b = \chi \, T_{ab} \qquad (56)$$

In fact $Q_a \cdot Q_b$ can be interpreted as spin-spin interaction energy:

$$Q_a \cdot Q_b / \chi = T_{ab} \qquad (57)$$

This means that where is torsion there is also curvature, and so the mass. It is clear that equation (56) represent self-interaction energy of the field Q but, because of the quantized values of spin, we can consider, as fundamental objects, the pairs of spins (with interaction $S_1 \cdot S_2$) to describe better what happens at the Planck scale in our picture of space-time and matter. As we will see, this lead us to new topics about the uncertainty relation between energy and time.

We take a volume V enough small inside a particle (a nucleon), because we need to consider a particle as an extended body (notice that $L_{Pl} \sim 10^{20}$ times smaller than the typical hadron dimension $r \sim 10^{-13}$cm). Using eqs.(56, 57) we can write for the energy density ε:

$$\varepsilon = \Delta E/\Delta V = Q_0 \cdot Q_0/\chi \propto L_{Pl}^2 \hbar c n^2 \qquad (58)$$

where $Q_0 \propto L_{Pl}^2 n$ and n is the number of spins per unit volume $n = N/\Delta V$. It follows that

$$\Delta E \, \Delta V \propto L_{Pl}^2 \hbar c N^2 \qquad (59)$$

or, if we consider the volume that contains only one spin ($n = 1/\Delta V$),

$$\Delta E \, \Delta V \approx L_{Pl}^2 \hbar c \qquad (60)$$

where ΔE is the energy inside ΔV.

Now if we take two interacting spins, separated from a distance Δd (where d is less than Compton length but greater than Planck Length), because L_{Pl} is the minimal length and the time Δt needed to connect them causally is $\Delta d/c$, then putting $\Delta V = L_{Pl}^2 c\Delta t$ in eq.(60) we find $\Delta E\Delta t \approx \hbar$ (remember that we are always in the situation where the spins are antiparallel), i.e. the well-known uncertainty relation. We give this interpretation: if the spins interact for a time Δt there is a fluctuation in energy ΔE that is "virtual"; if the interaction is stationary, we have inside ΔV an energy that can constitute a real particle (see eq.(60)). For example we consider the nucleon mass and suppose that its energy can be made by N parts ΔE which obey the eq.(60): putting $E = N \, \Delta E$ and $r^3 = N \, \Delta V$, we find $N \approx 10^{20}$ and $n = N/r^3 = 10^{59} cm^{-3}$. This value gives very high torsion ($Q = L_{Pl}^2 n$), but if we consider the wave equation for Q inside the nucleon as a whole, then occur spin fluctuations above mentioned for which $\langle Q \rangle = 0$ and $\langle Q^2 \rangle/\chi \approx \varepsilon \approx mc^2/r^3$. Note that N is just the ratio r/L_{Pl}: this seems in agreement with wave interpretation i.e. there is one spin for every wavelength L_{Pl} inside r.

In other words we can say that with spin fluctuations one can determine the mass of the nucleon through the number $N \approx 10^{20}$ which represents, geometrically, the ratio between Compton wave length and Planck length. As Q is confined inside the nucleon we can take as starting point the number $N \approx 10^{20}$ arriving in that way to the determination of the mass of a nucleon.

Work is in progress in considering the problem of spin fluctuations inside a nucleon in relation to quark's confinement.

APPENDIX A

We know that when we consider the early universe we have to deal both with elementary particle physics using quantum theory and with cosmology using general relativity; but general relativity is developed in real space-time while quantum theory needs a complex manifold. How can we conciliate general relativity with quantum theory? We think that the answer lies in a reformulation of Dirac's theory in terms of space-time geometric calculus without any complex number.

At this purpose we will consider the geometric algebra that, with the multivector concept and the interpretation of imaginary units as generator of rotations, places tensors and spinors on the same foot: both are described in a real space-time.

Without going in details we can define the geometric product as [6, 7]:

$$ab = a \cdot b + a \wedge b \qquad \text{(A.1)}$$

We will see that Hestenes space-time algebra automatically incorporates the geometric structure of space-time. This can be done introducing first of all the *outer product* $a \wedge b$ which is different from the usual cross product in the sense that it has magnitude $|a||b|\sin\vartheta$ and shares its skew property $a \wedge b = - b \wedge a$, but is not a scalar or a vector: it is a *directed area*, or *bivector*, oriented in the plane containing a and b. One can visualize the outer product as the area swept out by displacing a along b with the orientation given by traversing the so formed parallelogram first along a and after along b vector.

One can generalize this notion to products of objects with higher dimensionality or *grade* [6] in the sense that if the bivector $a \wedge b$, which has grade 2, is swept along another vector c of grade 1, one obtain the directed volume $(a \wedge b) \wedge c$ which is a trivector of grade 3. Thus we are led to the notion of multivector.

Considering for instance two dimensional space and taking two orthonormal basis vector σ_1 and σ_2 we have :

$$\sigma_1 \cdot \sigma_1 = \sigma_2 \cdot \sigma_2 = 1 \qquad \sigma_1 \cdot \sigma_2 = 0 \tag{A.2}$$

$$\sigma_1 \wedge \sigma_1 = \sigma_2 \wedge \sigma_2 = 0 \tag{A.3}$$

The outer product $\sigma_1 \wedge \sigma_2$ is a directed area. Then we have four independent basis vectors as:

$$\text{scalar } 1, \quad \text{vectors } \sigma_1, \sigma_2 \text{ and bivector } \sigma_1 \wedge \sigma_2 \tag{A.4}$$

and with these four basis elements we can form a multivector:

$$A \equiv a_0 1 + a_1\sigma_1 + a_2\sigma_2 + a_3\sigma_1 \wedge \sigma_2 \tag{A.5}$$

In order to define the multiplication AB between two multivector A and B (where $B \equiv b_0 1 + b_1\sigma_1 + b_2\sigma_2 + b_3\sigma_1 \wedge \sigma_2$) we have to remember how to multiply the four geometric basis elements:

$$\sigma_1^2 = \sigma_1\sigma_1 = \sigma_1 \cdot \sigma_1 + \sigma_1 \wedge \sigma_1 = 1 = \sigma_2^2 \tag{A.6}$$

and

$$\sigma_1\sigma_2 = \sigma_1 \cdot \sigma_2 + \sigma_1 \wedge \sigma_2 = \sigma_1 \wedge \sigma_2 = - \sigma_2 \wedge \sigma_1 = -\sigma_2\sigma_1 \tag{A.7}$$

from which (being the geometric product associative):

$$(\sigma_1\sigma_2)\sigma_1 = - \sigma_2\sigma_1\sigma_1 = - \sigma_2 \quad \text{and} \quad (\sigma_1\sigma_2)\sigma_2 = \sigma_1 \tag{A.8}$$

also

$$\sigma_1(\sigma_1\sigma_2) = \sigma_2 \quad \text{and} \quad \sigma_2(\sigma_1\sigma_2) = -\sigma_1 \tag{A.9}$$

Moreover

$$(\sigma_1 \wedge \sigma_2)^2 = \sigma_1\sigma_2\sigma_1\sigma_2 = -\sigma_1\sigma_1\sigma_2\sigma_2 = -1 \tag{A.10}$$

(being

$$\sigma_1\sigma_2 = \sigma_1 \cdot \sigma_2 + \sigma_1 \wedge \sigma_2 = \sigma_1 \wedge \sigma_2 = -\sigma_2 \wedge \sigma_1 = -\sigma_2\sigma_1 \tag{A.11}$$)

One of the result of the above derivation is that the bivector $\sigma_1 \wedge \sigma_2$ has the geometric effect of rotating the vectors σ_1 and σ_2 by 90^0 in its own plane: this property together with that shown in the relation (A.10), indicates that the bivector $\sigma_1 \wedge \sigma_2$ plays the role of the unit imaginary.

So the bivector $\sigma_1 \wedge \sigma_2$ is the unit of directed area but is also the generator of rotations in the plane.

Note that we will indicate the bivector $\sigma_1 \wedge \sigma_2 = \sigma_1\sigma_2$ simply with 'i'.

We can now consider these things in 3-space, adding a third orthonormal vector σ_3 to our basis. In this case we can form scalar 1, vectors σ_1, σ_2, σ_3, bivectors $\sigma_1\sigma_2$, $\sigma_2\sigma_3$, $\sigma_3\sigma_1$,

$$\text{trivector } \sigma_1\sigma_2\sigma_3 \text{ (volume element)} \tag{A.12}$$

and from these 8 objects we can define multivectors.

We do not go through calculation but we like to consider these relations:

$$(\sigma_1\sigma_2)\sigma_3 = \sigma_1\sigma_2\sigma_3 \ , \tag{A.13}$$

$$(\sigma_1\sigma_2\sigma_3)\sigma_k = \sigma_k(\sigma_1\sigma_2\sigma_3) \tag{A.14}$$

and

$$(\sigma_1\sigma_2\sigma_3)^2 = \sigma_1\sigma_2\sigma_3\sigma_1\sigma_2\sigma_3 = \sigma_1\sigma_2\sigma_1\sigma_2\sigma_3^2 = -1 \qquad (A.15)$$

Now it is easy to show that the trivector $\sigma_1\sigma_2\sigma_3$ can be written as

$$\sigma_1\sigma_2\sigma_3 = \sigma_1\wedge\sigma_2\wedge\sigma_3 = i \qquad (A.16)$$

The trivector $\sigma_1\sigma_2\sigma_3$ is the highest grade object in the 3-dimensional space and is the unit pseudoscalar for 3-space. i is a geometric object and is different from the imaginary unit $\sqrt{-1}$ which is a scalar (complex).

At this point we can notice a very important fact: multiplying this 3-vector by σ_3, σ_1, σ_2 we obtain respectively:

$$\begin{aligned}(\sigma_1\sigma_2\sigma_3)\sigma_3 &= \sigma_1\sigma_2 = i\sigma_3\\ (\sigma_1\sigma_2\sigma_3)\sigma_1 &= \sigma_2\sigma_3 = i\sigma_1\\ (\sigma_1\sigma_2\sigma_3)\sigma_2 &= \sigma_3\sigma_1 = i\sigma_2\end{aligned} \qquad (A.17)$$

which is the Pauli algebra.

The important fact is that the generator of rotations can be interpreted as spin. We do not go through the details of these considerations because we like to consider directly the spacetime algebra. In four dimensional space-time, from four orthonormal vectors as a basis, we can construct 16 geometrical elements, that is one scalar, four vectors, six bivectors, four trivectors and one pseudoscalar.

We can express these sixteen geometrical elements as

1	γ_μ	$\{\sigma_k, i\sigma_k\}$	$i\gamma_\mu$
1-scalar	4-vectors	6-bivectors	4-pseudovectors
i			
1-pseudoscalar			

(A.18)

where

$$\sigma_k = \gamma_k\gamma_0 \qquad (A.19)$$

and the unit pseudoscalar of spacetime is

$$i = \gamma_0\gamma_1\gamma_2\gamma_3 = \sigma_1\sigma_2\sigma_3 \qquad (A.20)$$

The algebra (A.18) is the spacetime algebra or real Dirac algebra. The even elements of the basis (A.18) coincide with Pauli algebra defined by (A.12) because of the relation (A.20) which in turn satisfies

$$\sigma_k \cdot \sigma_j = (1/2)(\sigma_k\sigma_j + \sigma_j\sigma_k) = \delta_{kj} , \qquad (A.21)$$

the requirement of a basis in three-dimensional Euclidean space. The Pauli algebra (A.12) is an even subalgebra of the Dirac algebra (A.12) with respect to the selection of the unit timelike vector γ_0 and is an eight-dimensional linear space of spinors.

We can show [6] that with this geometric algebra is possible, for instance, to write Dirac equation without any imaginary i.e.

$$\hbar\nabla\psi\gamma_2\gamma_1 - (e/c)A\psi = mc\psi\gamma_0 \qquad (A.22)$$

where ∇ is the four-dimensional generalization of the gradient operator which, taking into account of the metric, is

$$\nabla = \gamma^\mu\partial_\mu \quad , \quad \partial_\mu = \partial/\partial x^\mu \ , \quad A = A_\mu\gamma^\mu = A^\mu\gamma_\mu \qquad (A.23)$$

and ψ is connected with Dirac column spinor by $\Psi = \psi u$ being u the unit column spinor. The bivector $\gamma_2\gamma_1$ substitute the unit imaginary and $(\hbar/2)\gamma_2\gamma_1$ is identified with the spin.

APPENDIX B

In Dirac algebra $\mathcal{D}$ every trivector T is dual of some vector a:

$$T = ia \ . \tag{B.1}$$

Multiplying (geometrical product) on the left by 'i', we obtain from (B.1)

$$iT = - a \ . \tag{B.2}$$

So the dual of a trivector T is a unique vector - *a*. This establishes an isomorphism of the four-dimensional linear space of trivectors to the four-dimensional vector space. That is why trivectors of $\mathcal{D}$ are often called pseudoscalars.

Taking account of the above fact, we can notice that the commutation relation (16):

$$[Q, R^{\alpha}] = \gamma_2 \wedge \gamma_1 \ L_{Pl}^{-3} \tag{B.3}$$

is quite compatible with the commutation relation

$$[x, p] = j\hbar \ , \qquad (j = \sqrt{-1}\ ). \tag{B.4}$$

Because Q and R^{α} in (B.3) are respectively the torsion trivector and curvature trivector belonging to the four-dimensional linear space of trivectors, while x and p in (B.4) are both vectors belonging to the vector space which is isomorphic to the space of trivectors. It is important to note that the unit imaginary j is replaced by the unit bivector $\gamma_2 \wedge \gamma_1$ in (B.3), which plays the role of j in quantum theory.

APPENDIX C

The 'relative direction' of the vectors *a* and *b* is completely characterized by the geometric product

$$ab = a \cdot b + a \wedge b \ . \tag{C.1}$$

The resolved parts of the vector a along and perpendicular to the vector b can be obtained by right multiplication (geometric product) of (C.1) by b^{-1} as shown below:

$$abb^{-1} = a \cdot bb^{-1} + a \wedge bb^{-1} \ . \tag{C.2}$$

This can be written as

$$a = a_{\parallel} + a_{\perp} \ , \tag{C.3}$$

where

$$a_{\parallel} = a \cdot bb^{-1} = \text{the resolved part of } a \text{ along } b \tag{C.4}$$

$$a_{\perp} = a \wedge bb^{-1} = \text{the resolved part of } a \text{ perpendicular to } b \tag{C.5}$$

Eqs.(C.4) and (C.5) can be expressed as

$$a_{\parallel}b = a \cdot b = ba_{\parallel} \ , \tag{C.6}$$

$$a_{\perp}b = a \wedge b = -\ ba_{\perp} \ . \tag{C.7}$$

Next we generalize the above case for a multivector M of an arbitrary grade k which determines a k-dimensional vector space called M-space. The relative direction of M and some vector a is completely characterized by the geometric product

$$aM = a \cdot M + a \wedge M \ . \tag{C.8}$$

As in the earlier case, the vector a is uniquely resolved into a vector $a_{\parallel}$ in the M-space and a vector $a_{\perp}$ orthogonal to the M-space:

$$a = a_{\parallel} + a_{\perp} \ , \tag{C.9}$$

where

$$a_{\parallel} = a \cdot M\, M^{-1} \ , \qquad (C.10)$$

$$a_{\perp} = a \wedge M\, M^{-1} \ . \qquad (C.11)$$

Eqs.(C.10) and (C.11) can be expressed as

$$a_{\parallel}M = a \cdot M = (-1)^{k+1} M a_{\parallel} \ , \qquad (C.12)$$

$$a_{\perp}M = a \wedge M = (-1)^{k} M a_{\perp} \ . \qquad (C.13)$$

The vector $a_{\parallel}$ determined by eq. (C.10) is called the <u>projection</u> of the vector a into the M-space, while the vector $a_{\perp}$ determined by the eq. (C.11) is called the <u>rejection</u> of the vector a from the M-space.

In view of the above, the operator D_I in (25) operating on a multivector M (say) and described by the inner product

$$\nabla_a \cdot M + \gamma \cdot [\omega^a, M]$$

is called the interior covariant derivative. Likewise the operator D_E in (26) operating on M and described by the outer product

$$\nabla \wedge M + \gamma \wedge [\omega^a, M]$$

is called the exterior derivative. Thus one can write the covariant derivative operator D as (see eq. (C.9)):

$$D = D_I + D_E \ . \qquad (C.14)$$

References

[1] H.-J.Treder - Ann.der Physik 48, 497 (1991)

[2] V.de Sabbata - Il Nuovo Cimento 107A, 363 (1994)

[3] H.-J.Treder - Astr.Nachr.315,1 (1994)

[4] H.-H.v.Borzeszkowski and H.-J.Treder - "Classical Gravity and Quantum Matter Field" in "Quantum Gravity" ed.by P.G.Bergmann, V.de Sabbata and H.-J.Treder, World Sci.Singapore, pp.32-42 (1996)

[5] H.-H.v.Borzeszkowski and H.-J.Treder - "Torsion and the Weyl-Cartan Space Problem in Purely Affine Theory" in "Spin in Gravity" ed.by P.G.Bergmann, V.de Sabbata, G.T.Gillies, and P.I.Pronin World Sci.Singapore, pp.9-32 (1998);

[6] B.K.Datta, V.de Sabbata and L.Ronchetti - Il Nuovo Cimento 113B, 711 (1998)

[7] D.Hestenes - "Clifford Algebra to Geometric calculus" D.Reidel Publ.Co. Dordrecht, Holland 1982

[8] A.Lasenby, C.Doran and S.Gull - Found.Phys. 23,1295 (1993)

[9] C.Doran, D.Hestenes, F.Sommer and N.Von Acker - J.Math.Phys. 34, 3642 (1993)

[10] F.W.Hehl, J.Nitsch and J.Von der Heyde - in "General Relativity and Gravitation" Proceedings of Jena Conference p.329-355 (1982)

[11] R.T.Rauch - GRG 14, 331 (1982)

[12] R.T.Rauch - Phys.Rev.D25, 577 (1982)

[13] R.T.Rauch - Phys.Rev.D26, 931 (1982)

[14] J.A.Schouten - "Ricci-calculus" Springer-Verlag, Berlin 1954 p.141

Noncommutative Geometry for Pedestrians*

J. Madore
Laboratoire de Physique Théorique
Université de Paris-Sud, Bâtiment 211, F-91405 Orsay

Max-Planck-Institut für Physik
Föhringer Ring 6, D-80805 München

Abstract

A short historical review is made of some recent literature in the field of noncommutative geometry, especially the efforts to add a gravitational field to noncommutative models of space-time and to use it as an ultraviolet regulator. An extensive bibliography has been added containing reference to recent review articles as well as to part of the original literature.

LMU-TPW 99-11

*Lecure given at the International School of Gravitation, Erice: 16th Course: 'Classical and Quantum Non-Locality'.

1 Introduction

To control the divergences which from the very beginning had plagued quantum electrodynamics, Heisenberg already in the 1930's proposed to replace the space-time continuum by a lattice structure. A lattice however breaks Lorentz invariance and can hardly be considered as fundamental. It was Snyder [201, 202] who first had the idea of using a noncommutative structure at small length scales to introduce an effective cut-off in field theory similar to a lattice but at the same time maintaining Lorentz invariance. His suggestion came however just at the time when the renormalization program finally successfully became an effective if rather *ad hoc* prescription for predicting numbers from the theory of quantum electrodynamics and it was for the most part ignored. Some time later von Neumann introduced the term 'noncommutative geometry' to refer in general to a geometry in which an algebra of functions is replaced by a noncommutative algebra. As in the quantization of classical phase-space, coordinates are replaced by generators of the algebra [60]. Since these do not commute they cannot be simultaneously diagonalized and the space disappears. One can argue [148] that, just as Bohr cells replace classical-phase-space points, the appropriate intuitive notion to replace a 'point' is a Planck cell of dimension given by the Planck area. If a coherent description could be found for the structure of space-time which were pointless on small length scales, then the ultraviolet divergences of quantum field theory could be eliminated. In fact the elimination of these divergences is equivalent to coarse-graining the structure of space-time over small length scales; if an ultraviolet cut-off Λ is used then the theory does not see length scales smaller than Λ^{-1}. When a physicist calculates a Feynman diagram he is forced to place a cut-off Λ on the momentum variables in the integrands. This means that he renounces any interest in regions of space-time of volume less than Λ^{-4}. As Λ becomes larger and larger the forbidden region becomes smaller and smaller but it can never be made to vanish. There is a fundamental length scale, much larger than the Planck length, below which the notion of a point is of no practical importance. The simplest and most elegant, if certainly not the only, way of introducing such a scale in a Lorentz-invariant way is through the introduction of noncommuting space-time 'coordinates'.

As a simple illustration of how a 'space' can be 'discrete' in some sense and still covariant under the action of a continuous symmetry group one can consider the ordinary round 2-sphere, which has acting on it the rotational group SO_3. As a simple example of a lattice structure one can consider two points on the sphere, for example the north and south poles. One immediately notices of course that by choosing the two points one has broken the rotational invariance. It can be restored at the expense of commutativity. The set of functions on the two points can be identified with the algebra of diagonal 2×2 matrices, each of the two entries on the diagonal corresponding to a possible value of a function at one of the two points. Now an action of a group on the lattice is equivalent to an action of the group on the matrices and there can obviously be no non-trivial action of the group SO_3 on the algebra of diagonal 2×2 matrices. However if one extends the algebra to the noncommutative algebra of all 2×2 matrices one recovers the invariance. The two points, so to speak, have been smeared out over the surface of a sphere; they are replaced by two cells. An 'observable' is an hermitian 2×2 matrix and has therefore two real eigenvalues, which are its values on the two cells. Although what we have just done has nothing to do with Planck's constant it is similar to the procedure of replacing a classical spin which can take two values by a quantum spin of total spin 1/2. Only the latter is invariant under the rotation group. By replacing the spin 1/2 by arbitrary spin s one can describe a 'lattice structure' of $n = 2s + 1$ points in an SO_3-invariant manner. The algebra becomes then the algebra M_n of $n \times n$ complex matrices and there are n cells of area $2\pi k\!\!\!{}^-$ with

$$n \simeq \frac{\mathrm{Vol}(S^2)}{2\pi k\!\!\!{}^-}.$$

In general, a static, closed surface in a fuzzy space-time as we define it can only have a finite number of modes and will be described by some finite-dimensional algebra [90, 92, 94, 95, 96]. Graded extensions of some of these algebras have also been constructed [97, 98]. Although we are interested in a matrix version of surfaces primarily as a model of an eventual noncommutative theory of gravity they have a certain interest in other, closely related, domain of physics. We have seen, for example, that without the differential calculus the fuzzy sphere is basically just an approximation to a classical spin r by a quantum spin r with $\hbar$ in lieu of $k\!\!\!{}^-$. It has been extended in various directions under various names and for various reasons [17, 58, 105, 22]. In order to explain the finite entropy of a black hole it has been conjectured, for example by 't Hooft [207], that the horizon has a structure of a fuzzy 2-sphere since the latter has a finite number of 'points' and yet has an SO_3-invariant geometry. The horizon of a black hole might be a unique situation in which one can actually 'see' the cellular structure of space.

It is to be stressed that we shall here modify the structure of Minkowski space-time but maintain covariance under the action of the Poincaré group. A fuzzy space-time looks then like a solid which has a homogeneous distribution of dislocations but no disclinations. We can pursue this solid-state analogy and think of the ordinary Minkowski coordinates as macroscopic order parameters obtained by coarse-graining over scales less than the fundamental scale. They break down and must be replaced by elements of some noncommutative algebra when one considers phenomena on these scales. It might be argued that since we have made space-time 'noncommutative' we ought to do the same with the Poincaré group. This logic leads naturally to the notion of a q-deformed Poincaré (or Lorentz) group which act on a very particular noncommutative version of Minkowski space called q-Minkowski space [141, 142, 28, 10, 30]. The idea of a q-deformation goes back to Sylvester [200]. It was taken up later by Weyl [212] and Schwinger [197] to produce a finite version of quantum mechanics.

It has also been argued, for conceptual as well as practical, numerical reasons, that a lattice version of space-time or of space is quite satisfactory if one uses a random lattice structure or graph. The most widely used and successful modification of space-time is in fact what is called the lattice approximation. From this point of view the Lorentz group is a classical invariance group and is not valid at the microscopic level. Historically the first attempt to make a finite approximation to a curved manifold was due to Regge and this developed into what is now known as the Regge calculus. The idea is based on the fact that the Euler number of a surface can be expressed as an integral of the gaussian curvature. If one applies this to a flat cone with a smooth vertex then one finds a relation between the defect angle and the mean curvature of the vertex. The latter is encoded in the former. In recent years there has been a burst of activity in this direction, inspired by numerical and theoretical calculations of critical exponents of phase transitions on random surfaces. One chooses a random triangulation of a surface with triangles of constant fixed length, the lattice parameter. If a given point is the vertex of exactly six triangles then the curvature at the point is flat; if there are less than six the curvature is positive; it there are more than six the curvature is negative. Non-integer values of curvature appear through statistical fluctuation. Attempts have been made to generalize this idea to three dimensions using tetrahedra instead of triangles and indeed also to four dimensions, with euclidean signature. The main problem, apart from considerations of the physical relevance of a theory of euclidean gravity, is that of a proper identification of the curvature invariants as a combination of defect angles. On the other hand some authors have investigated random lattices from the point of view of noncommutative geometry. For an introduction to the lattice theory of gravity from these two different points of view we refer to the books by Ambjørn & Jonsson [5] and by Landi [136]. Compare also the loop-space approach to quantum gravity [11, 82, 7].

One typically replaces the four Minkowski coordinates x^μ by four generators q^μ of a noncommutative algebra which satisfy commutation relations of the form

$$[q^\mu, q^\nu] = i\bar{k} q^{\mu\nu}. \tag{1.1}$$

The parameter $\bar{k}$ is a fundamental area scale which we shall suppose to be of the order of the Planck area:

$$\bar{k} \simeq \mu_P^{-2} = G\hbar.$$

There is however no need for this assumption; the experimental bounds would be much larger. Equation (1.1) contains little information about the algebra. If the right-hand side does not vanish it states that at least some of the q^μ do not commute. It states also that it is possible to identify the original coordinates with the generators q^μ in the limit $\bar{k} \to 0$:

$$\lim_{\bar{k}\to 0} q^\mu = x^\mu. \tag{1.2}$$

For mathematical simplicity we shall suppose this to be the case although one could include a singular 'renormalization constant' Z and replace (1.2) by an equation of the form

$$\lim_{k\to 0} q^\mu = Z x^\mu. \tag{1.3}$$

If, as we shall argue, gravity acts as a universal regulator for ultraviolet divergences then one could reasonably expect the limit $k \to 0$ to be a singular limit.

Let $\mathcal{A}_k$ be the algebra generated in some sense by the elements q^μ. We shall be here working on a formal level so that one can think of $\mathcal{A}_k$ as an algebra of polynomials in the q^μ although we shall implicitly suppose that there are enough elements to generate smooth functions on space-time in the commutative limit. Since we have identified the generators as hermitian operators on some Hilbert space we can identify $\mathcal{A}_k$ as a subalgebra of the algebra of all operators on the Hilbert space. We have added the subscript k to underline the dependence on this parameter but of course the commutation relations (1.1) do not determine the structure of $\mathcal{A}_k$, We in fact conjecture that every possible gravitational field can be considered as the commutative limit of a noncommutative equivalent and that the latter is strongly restricted if not determined by the structure of the algebra $\mathcal{A}_k$. We must have then a large number of algebras $\mathcal{A}_k$ for each value of k.

Interest in Snyder's idea was revived much later when mathematicians, notably Connes [42] and Woronowicz [214, 215], succeeded in generalizing the notion of differential structure to noncommutative geometry. Just as it is possible to give many differential structures to a given topological space it is possible to define many differential calculi over a given algebra. We shall use the term 'noncommutative geometry' to mean 'noncommutative differential geometry' in the sense of Connes. Along with the introduction of a generalized integral [50] this permits one in principle to define the action of a Yang-Mills field on a large class of noncommutative geometries.

One of the more obvious applications was to the study of a modified form of Kaluza-Klein theory in which the hidden dimensions were replaced by noncommutative structures [145, 146, 67]. In simple models gravity could also be defined [146, 147] although it was not until much later [171, 69, 117] that the technical problems involved in the definition of this field were to be to a certain extent overcome. Soon even a formulation of the standard model of the electroweak forces could be given [48]. A simultaneous development was a revival [161, 52, 145] of the idea of Snyder that geometry at the Planck scale would not necessarily be described by a differential manifold.

One of the advantages of noncommutative geometry is that smooth, finite examples [148] can be constructed which are invariant under the action of a continuous symmetry group. Such models necessarily have a minimal length associated to them and quantum field theory on them is necessarily finite [90, 92, 94, 24]. In general this minimal length is usually considered to be in some way or another associated with the gravitational field. The possibility which we shall consider here is that the mechanism by which this works is through the introduction of noncommuting 'coordinates'. This idea has been developed by several authors [103, 148, 62, 124, 73, 123, 31] from several points of view since the original work of Snyder. It is the left-hand arrow of the diagram

$$\begin{array}{ccc} \mathcal{A}_{\bar{k}} & \Longleftarrow & \Omega^*(\mathcal{A}_{\bar{k}}) \\ \Downarrow & & \Uparrow \\ \text{Cut-off} & & \text{Gravity} \end{array} \tag{1.4}$$

The $\mathcal{A}_{\bar{k}}$ is a noncommutative algebra and the index $\bar{k}$ indicates the area scale below which the noncommutativity is relevant; this would normally be taken to be the Planck area.

The top arrow is a mathematical triviality; the $\Omega^*(\mathcal{A}_{\bar{k}})$ is a second algebra which contains $\mathcal{A}_{\bar{k}}$ and is what gives a differential structure to it just as the algebra of de Rham differential forms gives a differential structure to a smooth manifold. There is an associated differential d, which satisfies the relation $d^2 = 0$. The couple $(\Omega^*(\mathcal{A}), d)$ is known as a differential calculus over the algebra $\mathcal{A}$. The algebra $\mathcal{A}$ is what in ordinary geometry would determine the set of points one is considering, with possibly an additional topological or measure theoretic structure. The differential calculus is what gives an additional differential structure or a notion of smoothness. On a commutative algebra of functions on a lattice, for example, it would determine the number of nearest neighbours and therefore the dimension. The idea of extending the notion of a differential to noncommutative algebras is due to Connes [42, 45, 48, 49] who proposed a definition based on a formal analogy with an identity in ordinary geometry involving the Dirac operator $\not{D}$. Let ψ be a Dirac spinor and f a smooth function. Then one can write

$$i\gamma^\alpha e_\alpha f\psi = \not{D}(f\psi) - f\not{D}\psi.$$

Here e_α is the Pfaffian derivative with respect to an orthonormal moving frame θ^α. This equation can be written

$$\gamma^\alpha e_\alpha f = -i[\not{D}, f]$$

and it is clear that if one makes the replacement

$$\gamma^\alpha \mapsto \theta^\alpha$$

then on the right-hand side one has the de Rham differential. Inspired by this fact, one defines a differential in the noncommutative case by the formula

$$df = i[F, f]$$

where now f belongs to a noncommutative algebra $\mathcal{A}$ with a representation on a Hilbert space $\mathcal{H}$ and F is an operator on $\mathcal{H}$ with spectral properties which make it look like a Dirac operator. The triple $(\mathcal{A}, F, \mathcal{H})$ is called a spectral triple. It is inspired by the K-cycle introduced by Atiyah [9] to define a dual to K-theory [8]. The simplest example is obtained by choosing $\mathcal{A} = \mathbb{C} \oplus \mathbb{C}$ acting on $\mathbb{C}^2$ by left multiplication and

$$F = \begin{pmatrix} 0 & 1 \\ 1 & 0 \end{pmatrix}.$$

The 1-forms are then off-diagonal 2×2 complex matrices. The differential is extended to them using the same formula as above but with a bracket which is an anticommutator instead of a commutator. Since $F^2 = 1$ it is immediate that $d^2 = 0$. The algebra $\mathcal{A}$ of this example can be considered as the algebra of functions on 2 points and the differential can be identified with the finite-difference operator.

One can argue [59, 156, 152], not completely successfully, that each gravitational field is the unique 'shadow' in the limit $\hbar \to 0$ of some differential structure over some noncommutative algebra. This would define the right-hand arrow of the diagram. A hand-waving argument can be given [21, 154] which allows one to think of the noncommutative structure of space-time as being due to quantum fluctuations of the light-cone in ordinary 4-dimensional space-time. This relies on the existence of quantum gravitational fluctuations. A purely classical argument based on the formation of black-holes has been also given [62]. In both cases the classical gravitational field is to be considered as regularizing the ultraviolet divergences through the introduction of the noncommutative structure of space-time. This can be strengthened as the conjecture that the classical gravitational field and the noncommutative nature of space-time are two aspects of the same thing. If the gravitational field is quantized then presumably the light-cone will fluctuate and any two points with a space-like separation would have a time-like separation on a time scale of the order of the Planck time, in which case the corresponding operators would no longer commute. So even in flat space-time quantum fluctuations of the gravitational field could be expected to introduce a non-locality in the theory. This is one possible source of noncommutative geometry on the order of the Planck scale. The composition of the three arrows in (1.4) is an expression of an old idea, due to Pauli, that perturbative ultraviolet divergences will somehow be regularized by the gravitational field [57, 107]. We refer to Garay [84] for a recent review.

One example from which one can seek inspiration in looking for examples of noncommutative geometries is quantized phase space, which had been already studied from a noncommutative point of view by Dirac [60]. The minimal length in this case is given by the Heisenberg uncertainty relations or by modifications thereof [124]. In fact in order to explain the supposed Zitterbewegung of the electron Schrödinger [193] had proposed to mix position space with momentum space in order to obtain a set of center-of-mass coordinates which did not commute. This idea has inspired many of the recent attempts to introduce minimal lengths. We refer to [73, 123] for examples which are in one way or another connected to noncommutative geometry. Another concept from quantum mechanics which is useful in concrete applications is that of a coherent state. This was first used in a finite noncommutative geometry by Grosse & Prešnajder [91] and later applied [123, 33, 39] to the calculation of propagators on infinite noncommutative geometries, which now become regular 2-point functions and yield finite vacuum fluctuations. Although efforts have been made in this direction [39] these fluctuations have not been satisfactorily included as a source of the gravitational field, even in some 'quasi-commutative' approximation. If this were done then the missing arrow in (1.4) could be drawn. The difficulty is partly due to the lack of tractable noncommutative versions of curved spaces.

The fundamental open problem of the noncommutative theory of gravity concerns of course the relation it might have to a future quantum theory of gravity either directly or via the theory of 'strings' and 'membranes'. But there are more immediate technical problems which have not received a satisfactory answer. We shall mention the problem of the definition of the curvature. It is not certain that the ordinary definition of curvature taken directly from differential geometry is the quantity which is most useful in the noncommutative theory. Cyclic homology groups have been proposed by Connes as the appropriate generalization to noncommutative geometry of topological invariants; the definition of other, non-topological, invariants in not clear. It is not in fact even obvious that one should attempt to define curvature invariants.

There is an interesting theory of gravity, due to Sakharov and popularized by Wheeler, called induced gravity, in which the gravitational field is a phenomenological coarse-graining of more fundamental fields. Flat Minkowski space-time is to be considered as a sort of perfect crystal and curvature as a manifestation of elastic tension, or possibly of defects, in this structure. A deformation in the crystal produces a variation in the vacuum energy which we perceive as gravitational energy. 'Gravitation is to particle physics as elasticity is to chemical physics: merely a statistical measure of residual energies.' The description of the gravitational field which we are attempting to formulate using noncommutative geometry is not far from this. We have noticed that the use of noncommuting coordinates is a convenient way of making a discrete structure like a lattice invariant under the action of a continuous group. In this sense what we would like to propose is a Lorentz-invariant version of Sakharov's crystal. Each coordinate can be separately measured and found to have a distribution of eigenvalues similar to the distribution of atoms in a crystal. The gravitational field is to be considered as a measure of the variation of this distribution just as elastic energy is a measure of the variation in the density of atoms in a crystal.

We shall here accept a noncommutative structure of space-time as a mathemetical possibility. One can however attempt to associate the structure with other phenomena. A first step in this direction was undoubtedly taken by Bohr & Rosenfeld [21] when they deduced an intrinsic uncertainty in the position of an event in space-time from the quantum-mechanical measurement process. This idea has been since pursued by other authors [4] and even related to the formation of black holes [62, 137] and to the influence of quantum fluctuations in the gravitational field [154, 7]. An uncertainty relation in the measurements of an event is one of the most essential aspects of a noncommutative structure. The possible influence of quantum-mechanical fluctuations on differential forms was realized some time ago by Segal [198]. A related idea is what one might refer to as 'spontaneous lattization'. A quantum operator is a very singular object in general and the correct definition of the space-time coordinates, considered as quantum operators, could give rise to a preferred set of events in space-time which has some of the aspects of a 'lattice' in the sense that each operator, has a discrete spectrum [196, 124, 73, 123, 122]. The work of Yukawa [218] and Takano [205] could be considered as somewhat similar to this, except that the fuzzy nature of space-time is emphasized and related to the presence of particles. Finkelstein [75] has attempted a very philosophical derivation of the structure of space-time from the notion of 'simplicity' (in the group-theoretic sense of the word) which has led him to the possibility of the 'superposition of points', simething very similar to noncommutativity. We shall mention below the attempts to derive a noncommutative structure of space-time from string theory.

When referring to the version of space-time which we describe here we use the adjective 'fuzzy' to underline the fact that points are ill-defined. Since the algebraic structure is described by commutation relations the qualifier 'quantum' has also been used [201, 62, 156]. This latter expression is unfortunate since the structure has no immediate relation to quantum mechanics and also it leads to confusion with 'spaces' on which 'quantum groups' act. To add to the confusion the word 'quantum' has also been used [87] to designate equivalence classes of ordinary differential geometries which yield isomorphic string theories and the word 'lattice' has been used [201, 73, 207] to designate what we here qualify as 'fuzzy'.

2 A simple example

The algebra $\mathcal{P}(u, v)$ of polynomials in $u = e^{ix}$, $v = e^{iy}$ is dense in any algebra of functions on the torus, defined by the relations $0 \leq x \leq 2\pi$, $0 \leq y \leq 2\pi$, where x and y are the ordinary cartesian coordinates of $\mathbb{R}^2$. If one considers a square lattice of n^2 points then $u^n = 1$ and $v^n = 1$ and the algebra is reduced to a subalgebra $\mathcal{P}_n$ of dimension n^2. Introduce a basis $|j\rangle_1$, $0 \leq j \leq n-1$, of $\mathbb{C}^n$ with $|n\rangle_1 \equiv |0\rangle_1$ and replace u and v by the operators

$$u|j\rangle_1 = q^j|j\rangle_1, \quad v|j\rangle_1 = |j+1\rangle_1, \qquad q^n = 1.$$

Then the new elements u and v satisfy the relations

$$uv = qvu, \quad u^n = 1, \qquad v^n = 1$$

and the algebra they generate is the matrix algebra M_n instead of the commutative algebra $\mathcal{P}_n$. There is also a basis $|j\rangle_2$ in which v is diagonal and a 'Fourier' transformation between the two [197].

Introduce the forms [157]

$$\theta^1 = -i\Big(1 - \frac{n}{n-1}|0\rangle_2\langle 0|\Big)u^{-1}du,$$
$$\theta^2 = -i\Big(1 - \frac{n}{n-1}|n-1\rangle_1\langle n-1|\Big)v^{-1}dv.$$

In this simple example the differential calculus can be defined by the relations

$$\theta^a f = f\theta^a, \qquad \theta^a\theta^b = -\theta^b\theta^a$$

of ordinary differential geometry. It follows that

$$\Omega^1(M_n) \simeq \bigoplus_1^2 M_n, \qquad d\theta^a = 0.$$

The differential calculus has the form one might expect of a noncommutative version of the torus. Notice that the differentials du and dv do not commute with the elements of the algebra.

One can choose for q the value

$$q = e^{2\pi i l/n}$$

for some integer l relatively prime with respect to n. The limit of the sequence of algebras as $l/n \to \alpha$ irrational is known as the rotation algebra or the noncommutative torus [184]. This algebra has a very rich representation theory and it has played an important role as an example in the developement of noncommutative geometry [50].

3 Noncommutative electromagnetic theory

The group of unitary elements of the algebra of functions on a manifold is the local gauge group of electromagnetism and the covariant derivative associated to the electromagnetic potential can be expressed as a map

$$\mathcal{H} \xrightarrow{D} \Omega^1(V) \otimes_{\mathcal{A}} \mathcal{H} \tag{3.1}$$

from a $\mathcal{C}(V)$-module $\mathcal{H}$ to the tensor product $\Omega^1(V) \otimes_{\mathcal{C}(V)} \mathcal{H}$, which satisfies a Leibniz rule

$$D(f\psi) = df \otimes \psi + fD\psi, \qquad f \in \mathcal{C}(V), \quad \psi \in \mathcal{H}.$$

We shall often omit the tensor-product symbol in the following. As far as the electromagnetic potential is concerned we can identify $\mathcal{H}$ with $\mathcal{C}(V)$ itself;

electromagnetism couples equally, for example, to all four components of a Dirac spinor. The covariant derivative is defined therefore by the Leibniz rule and the definition

$$D1 = A \otimes 1 = A.$$

That is, one can rewrite (3.1) as

$$D\psi = (\partial_\mu + A_\mu)dx^\mu \psi.$$

One can study electromagnetism on a large class of noncommutative geometries [146, 67, 48, 53] and there exist many recent reviews [209, 152, 116]. Because of the noncommutativity however the result often looks more like non-abelian Yang-Mills theory.

4 Metrics

We shall define a metric as a bilinear map

$$\Omega^1(\mathcal{A}) \otimes_{\mathcal{A}} \Omega^1(\mathcal{A}) \xrightarrow{g} \mathcal{A}. \tag{4.1}$$

This is a 'conservative' definition, a straightforward generalization of one of the possible definitions of a metric in ordinary differential geometry:

$$g(dx^\mu \otimes dx^\nu) = g^{\mu\nu}.$$

The usual definition of a metric in the commutative case is a bilinear map

$$\mathcal{X} \otimes_{\mathcal{C}(V)} \mathcal{X} \xrightarrow{g} \mathcal{C}(V)$$

where $\mathcal{X}$ is the $\mathcal{C}(V)$-bimodule of vector fields on V:

$$g(\partial_\mu \otimes \partial_\nu) = g_{\mu\nu}.$$

This definition is not suitable in the noncommutative case since the set of derivations of the algebra, which is the generalization of $\mathcal{X}$, has no natural structure as an $\mathcal{A}$-module. The linearity condition is equivalent to a locality condition for the metric; the length of a vector at a given point depends only on the value of the metric and the vector field at that point. In the noncommutative case bilinearity is the natural (and only possible) expression of locality. It would exclude, for example, a metric in ordinary geometry defined by a map of the form

$$g(\alpha, \beta)(x) = \int_V g_x(\alpha_x, \beta_y) G(x, y) dy.$$

Here $\alpha, \beta \in \Omega^1(V)$ and g_x is a metric on the tangent space at the point $x \in V$. The function $G(x, y)$ is an arbitrary smooth function of x and y and dy is the measure on V induced by the metric.

Introduce a bilinear flip σ:

$$\Omega^1(\mathcal{A}) \otimes_{\mathcal{A}} \Omega^1(\mathcal{A}) \xrightarrow{\sigma} \Omega^1(\mathcal{A}) \otimes_{\mathcal{A}} \Omega^1(\mathcal{A}) \tag{4.2}$$

We shall say that the metric is symmetric if

$$g \circ \sigma \propto g.$$

Many of the finite examples have unique metrics [158] as do some of the infinite ones [31]. Other definitions of a metric have been given, some of which are similar to that given above but which weaken the locality condition [32] and one [49] which defines a metric on the associated space of states.

5 Linear Connections

An important geometric problem is that of comparing vectors and forms defined at two different points of a manifold. The solution to this problem leads to the concepts of a connection and covariant derivative. We define a linear connection as a covariant derivative

$$\Omega^1(\mathcal{A}) \xrightarrow{D} \Omega^1(\mathcal{A}) \otimes_{\mathcal{A}} \Omega^1(\mathcal{A})$$

on the $\mathcal{A}$-bimodule $\Omega^1(\mathcal{A})$ with an extra right Leibniz rule

$$D(\xi f) = \sigma(\xi \otimes df) + (D\xi) f$$

defined using the flip σ introduced in (4.2). In ordinary geometry the map

$$D(dx^\lambda) = -\Gamma^\lambda_{\mu\nu} dx^\mu \otimes dx^\nu$$

defines the Christophel symbols.

We define the torsion map

$$\Theta : \Omega^1(\mathcal{A}) \to \Omega^2(\mathcal{A})$$

by $\Theta = d - \pi \circ D$. It is left-linear. A short calculation yields

$$\Theta(\xi) f - \Theta(\xi f) = \pi \circ (1 + \sigma)(\xi \otimes df).$$

We shall impose the condition

$$\pi \circ (\sigma + 1) = 0 \tag{5.1}$$

on σ. It could also be considered as a condition on the product π. In fact in ordinary geometry it is the definition of π; a 2-form can be considered as an antisymmetric tensor. Because of this condition the torsion is a bilinear map. Using σ a reality condition on the metric and the linear connection can be introduced [78]. In the commutative limit, when it exists, the commutator defines a Poisson structure, which normally would be expected to have an intimate relation with the linear connection. This relation has only been studied in very particular situations [149].

6 Gravity

The classical gravitational field is normally supposed to be described by a torsion-free, metric-compatible linear connection on a smooth manifold. One might suppose that it is possible to formulate a noncommutative theory of (classical/quantum) gravity by replacing the algebra of functions by a more general algebra and by choosing an appropriate differential calculus. It seems however difficult to introduce a satisfactory definition of local curvature and the corresponding curvature invariants [55, 68, 56]. One way of circumventing this problem is to consider classical gravity as an effective theory and the Einstein-Hilbert action as an induced action. We recall that the classical gravitational action is given by

$$S[g] = \mu_P^4 \Lambda_c + \mu_P^2 \int R.$$

In the noncommutative case there is a natural definition of the integral [50, 43, 45] but there does not seem to be a natural generalization of the Ricci scalar. One of the problems is the fact that the natural generalization of the curvature form is in general not right-linear in the noncommutative case. The Ricci scalar then will not be local. One way of circumventing these problems is to return to an old version of classical gravity known as induced gravity [185, 186]. The idea is to identify the gravitational action with the quantum corrections to a classical field in a curved background. If $\Delta[g]$ is the operator which describes the propagation of a given mode in presence of a metric g then one finds that, with a cut-off Λ, the effective action is given by

$$\Gamma[g] \propto \operatorname{Tr} \log \Delta[g] \simeq \Lambda^4 \mathrm{Vol}(V)[g] + \Lambda^2 S_1[g] + (\log \Lambda) S_2[g] + \cdots .$$

If one identifies $\Lambda = \mu_P$ then one finds that $S_1[g]$ is the Einstein-Hilbert action. A problem with this is that it can be only properly defined on a compact manifold with a metric of euclidean signature and Wick rotation on a curved space-time is a rather delicate if not dubious procedure. Another problem with this theory, as indeed with the gravitational field in general, is that it predicts an extremely large cosmological constant. The expression $\operatorname{Tr} \log \Delta[g]$ has a natural generalization to the noncommutative case [111, 1, 34].

We have defined gravity using a linear connection, which required the full bimodule structure of the $\mathcal{A}$-module of 1-forms. One can argue that this was necessary to obtain a satisfactory definition of locality as well as a reality condition. It is possible to relax these requirements and define gravity as a Yang-Mills field [35, 135, 36, 81] or as a couple of left and right connections [55, 56]. If the algebra is commutative (but not an algebra of smooth functions) then to a certain extent all definitions coincide [136, 12].

7 Regularization

Using the diagram (1.4) we have argued that gravity regularizes propagators in quantum field theory through the formation of a noncommutative structure. Several explicit examples of this have been given in the literature [202, 148, 62, 124, 129, 39]. In particular an energy-momentum tensor constructed from regularized propagators [39] has been used as a source of a cosmological solution. The propagators appear as if they were derived from non-local theories on ordinary space-time [218, 175, 119, 205]. We required that the metric that we use be local in the sense that the map (4.1) is bilinear with respect to the algebra. One could say that the theory is as local as the algebra will permit. However, since the algebra is not an algebra of points this means that the theory appears to be non-local as an effective theory on a space-time manifold.

8 Kaluza-Klein theory

We mentioned in the Introduction that one of the first, obvious applications of noncommutative geometry is as an alternative hidden structure of Kaluza-Klein theory. This means that one leaves space-time as it is and one modifies only the extra dimensions; one replaces their algebra of functions by a noncommutative algebra, usually of finite dimension to avoid the infinite tower of massive states of traditional Kaluza-Klein theory. Because of this restriction and because the extra dimensions are purely algebraic in nature the length scale associated with them can be arbitrary [153], indeed as large as the Compton wave length of a typical massive particle.

The algebra of Kaluza-Klein theory is therefore, for example, a product algebra of the form

$$\mathcal{A} = \mathcal{C}(V) \otimes M_n.$$

Normally V would be chosen to be a manifold of dimension four, but since much of the formalism is identical to that of the M(atrix)-theory of D-branes [14, 83, 51]. For the simple models with a matrix extension one can use as gravitational action the Einstein-Hilbert action in 'dimension' $4+d$, including possibly Gauss-Bonnet terms [147, 153, 152, 154, 121]. For a more detailed review we refer to a lecture [155] at the 5th Hellenic school in Corfu.

9 Quantum groups and spaces

The set of smooth functions on a manifold is an algebra. This means that from any function of two variables one can construct a function of one by multiplication. If the manifold happens to be a Lie group then there is another operation which to any function of one variable constructs a function of two.

This is called co-multiplication and is usually written Δ:

$$(\Delta f)(g_1, g_2) = f(g_1 g_2).$$

It satisfies a set of consistency conditions with the product. Since the expression 'noncommutative group' designates something else the noncommutative version of an algebra of smooth functions on a Lie group has been called a 'quantum group'. It is neither 'quantum' nor 'group'. The first example was found by Kulish & Reshetikhin [133] and by Sklyanin [199]. A systematic description was first made by Woronowicz [213], by Jimbo [109], Manin [167, 168] and Drinfeld [65]. The Lie group $SO(n)$ acts on the space $\mathbb{R}^n$; the Lie group $SU(n)$ acts on $\mathbb{C}^n$. The 'quantum' versions $SO_q(n)$ and $SU_q(n)$ of these groups act on the 'quantum spaces' $\mathbb{R}^n_q$ and $\mathbb{C}^n_q$. These latter are noncommutative algebras with special covariance properties. The first differential calculus on a quantum space was constructed by Wess & Zumino [211]. There is an immense literature on quantum groups and spaces, from the algebraic as will as geometric point of view. We have included some of it in the bibliography. We mention in particular the collection of articles edited by Doebner & Hennig [61] and Kulish [130] and the introductory text by Kassel [113].

10 Mathematics

At a more sophisticated level one would have to add a topology to the algebra. Since we have identified the generators as hermitian operators on a Hilbert space, the most obvious structure would be that of a von Neumann algebra. We refer to Connes [45] for a description of these algebras within the context of noncommutative geometry. A large part of the interest of mathematicians in noncommutative geometry has been concerned with the generalization of topological invariants [42, 55, 170] to the noncommutative case. It was indeed this which lead Connes to develop cyclic cohomology. Connes [50, 43] has also developed and extended the notion of a Dixmier trace on certain types of algebras as a possible generalization of the notion of an integral. The representation theory of quantum groups is an active field of current interest since the pioneering work of Woronowicz [216]. For a recent survey we refer to the book by Klymik & Schmüdgen [125]. Another interesting problem is the relation between differential calculi covariant under the (co-)action of quantum groups and those constructed using the spectral-triple formalism of Connes. Although it has been known for some time [68, 59, 86, 41, 79] that many if not all of the covariant calculi have formal Dirac 'operators' it is only recently that mathematicians have considered [190, 191] to what extent these 'operators' can be actually represented as real operators on a Hilbert space and to what extent they satisfy the spectral-triple conditions.

11 String Theory

Last, but not least, is the possible relation of noncommutative geometry to string theory. We have mentioned that since noncommutative geometry is pointless a field theory on it will be divergence-free. In particular monopole configurations will have finite energy, provided of course that the geometry in which they are constructed can be approximated by a noncommutative geometry, since the point on which they are localized has been replaced by an volume of fuzz, This is one characteristic that it shares with string theory. Certain monopole solutions in string theory have finite energy [85] since the point in space (a D-brane) on which they are localized has been replaced by a throat to another 'adjacent' D-brane.

In noncommutative geometry the string is replaced by a certain finite number of elementary volumes of 'fuzz', each of which can contain one quantum mode. Because of the nontrivial commutation relations the 'line' $\delta q^\mu = q^{\mu\prime} - q^\mu$ joining two points $q^{\mu\prime}$ and q^μ is quantized and can be characterized [39] by a certain number of creation operators a_j each of which creates a longitudinal displacement. They would correspond to the rigid longitudinal vibrational modes of the string. Since it requires no energy to separate two points the string tension would be zero. This has not much in common with traditional string theory.

We mentioned in the previous section that noncommutative Kaluza-Klein theory has much in common with the M(atrix) theory of D-branes. What is lacking is a satisfactory supersymmetric extension. Finally we mention that there have been speculations that string theory might give rise naturally to space-time uncertainty relations [137] and that it might also give rise [108] to a noncommutative theory of gravity. More specifically there have been attempts [64, 63, 192] to relate a noncommutative structure of space-time to the quantization of the open string in the presence of a non-vanishing B-field.

Acknowledgments

The author would like to thank the Max-Planck-Institut für Physik in München for financial support and J. Wess for his hospitality there.

What follows constitutes in no way a complete bibliography of noncommutative geometry. It is strongly biased in favour of the author's personal interests and the few subjects which were touched upon in the text.

References

[1] T. Ackermann, J. Tolksdorf, "A generalized Lichnerowicz formula, the Wodzicki Residue and Gravity", *J. Geom. Phys.* **19** (1996) 143.

[2] A. Aghamohammadi, "The two-parametric extension of h deformation of $GL(2)$ and the differential calculus on its quantum plane", *Mod. Phys. Lett. A* **8** (1993) 2607.

[3] A. Aghamohammadi, M. Khorrami, A. Shariati, "$SL_h(2)$-Symmetric Torsionless Connections", *Lett. Math. Phys.* **40** (1997) 95.

[4] D.V. Ahluwalia, "Quantum Measurement, Gravitation and Locality", *Phys. Lett.* **B339** (1994) 301.

[5] J. Ambjørn, T. Jonsson, *Quantum Geometry, a statistical field theory approach*, Cambridge University Press, 1997.

[6] P. Aschieri, L. Castellani, "Bicovariant calculus on twisted $ISO(N)$, Quantum Poincaré Group and Quantum Minkowski Space", *Int. J. Mod. Phys. A* **11** (1996) 4513.

[7] A. Ashtekar, A. Corichi, J.A. Zapata, "Quantum Theory of Geometry III: Non-commutativity of Riemannian Structures", *Class. and Quant. Grav.* **15** (1998) 2955.

[8] M. Atiyah, *K-Theory*, W.A. Benjamin, New York, 1967.

[9] M. Atiyah, *Global theory of elliptic operators*, Proceedings of the Int. Symp. on Funct. Analysis, Tokyo, 21, 1969.

[10] J.A. de Azcárraga, P.P. Kulish, F. Rodenas, "Twisted h-spacetimes and invariant equations", *Z. Physik C - Particles and Fields* **76** (1997) 567.

[11] J.C. Baez, J.P. Muniain, *Gauge Fields, Knots, and Gravity*, Series on Knots and Everything No. 4, World Scientific Publishing, 1994.

[12] A.P. Balachandran, G. Bimonte, G. Landi, F. Lizzi, P. Teotonio-Sobrinho, "Lattice Gauge Fields and Noncommutative Geometry", *J. Geom. Phys.* **24** (1998) 353.

[13] B.S. Balakrishna, F. Gürsey, K.C. Wali, "Towards a Unified Treatment of Yang-Mills and Higgs Fields", *Phys. Rev.* **D44** (1991) 3313.

[14] T. Banks, W. Fischler, S.H. Shenker, L. Susskind, "M Theory as a Matrix Model: a Conjecture", *Phys. Rev.* **D55** (1997) 5112.

[15] P. Baum, R.G. Douglas, "K-Homology and Index Theory", *Proc. Symp. Pure Math.* **38** (1982) 117.

[16] J. Bellissard, A. van Elst, H. Schulz-Baldes, "The noncommutative geometry of the quantum Hall effect", *J. Math. Phys.* **35** (1994) 5373.

[17] F. Berezin, "General Concept of Quantization", *Comm. Math. Phys.* **40** (1975) 153.

[18] L.C. Biedenharn, "The quantum group $SU_q(2)$ and a q-analogue of the boson operators", *J. Phys. A: Math. Gen.* **22** (1989) L873.

[19] G. Bimonte, F. Lizzi, G. Sparano, "Distances on a Lattice from Noncommutative Geometry", *Phys. Lett.* **B341** (1994) 139.

[20] G. Bimonte, E. Ercolessi, G. Landi, F. Lizzi, G. Sparano, "Lattices and their Continuum Limits", *J. Math. Phys.* **20** (1996) 318.

[21] N. Bohr, L. Rosenfeld, "Zur Frage der Meßbarkeit der elektromagnetischen Feldgrößen", *Kgl. Danske. Vidensk. Selskab. Mat-fyz. Meddelser* **12** (1933) No. 8.

[22] M. Bordemann, J. Hoppe, P. Schaller, M. Schlichenmaier, "$gl(\infty)$ and Geometric Quantization", *Comm. Math. Phys.* **138** (1991) 209.

[23] L. Carminati, B. Iochum, D. Kastler, T. Schücker, *On Connes' New Principle of General Relativity. Can Spinors Hear the Forces of Space-Time?*, Operator Algebras and Quantum Field Theory, International Press, 1997.

[24] U. Carow-Watamura, S. Watamura, "Chirality and Dirac Operator on Noncommutative Sphere", *Comm. Math. Phys.* **183** (1997) 365.

[25] U. Carow-Watamura, M. Schlieker, M. Scholl, S. Watamura, "Tensor representations of the quantum group $SL_q(2, \mathbb{C})$ and quantum Minkowksi space", *Z. Physik C - Particles and Fields* **48** (1990) 159.

[26] U. Carow-Watamura, M. Schlieker, S. Watamura, M. Weich, "Bicovariant differential calculus on quantum groups $SU_q(N)$ and $SO_q(N)$", *Comm. Math. Phys.* **142** (1991) 605.

[27] U. Carow-Watamura, M. Schlieker, M. Scholl, S. Watamura, "Quantum Lorenz Group", *Int. J. Mod. Phys. A* **6** (1991) 3081.

[28] L. Castellani, "Differential Calculus on $ISO_q(N)$, Quantum Poincaré algebra and q-Gravity", *Comm. Math. Phys.* **171** (1995) 383.

[29] L. Castellani, R. D'Auria, P. Fré, *Supergravity and Superstrings, a Geometric Perspective*, World Scientific Publishing, 1991.

[30] B.L. Cerchiai, J. Wess, "q-Deformed Minkowski Space based on a q-Lorentz Algebra", *Euro. Phys. J. C* **5** (1998) 553.

[31] B.L. Cerchiai, R. Hinterding, J. Madore, J. Wess, "The Geometry of a q-deformed Phase Space", *Euro. Phys. J. C* **8** (1999) 533.

[32] B.L. Cerchiai, R. Hinterding, J. Madore, J. Wess, "A Calculus Based on a q-deformed Heisenberg Algebra", *Euro. Phys. J. C* **8** (1999) 547.

[33] M. Chaichian, A. Demichev, P. Prešnajder, "Quantum Field Theory on Noncommutative Space-Times and the Persistence of Ultraviolet Divergences", Helsinki Preprint HIP-1998-77/TH, hep-th/9812180.

[34] A.H. Chamseddine, A. Connes, "Universal Formula for Noncommutative Geometry Actions: Unification of Gravity and the Standard Model", *Phys. Rev. Lett.* **77** (1996) 4868.

[35] A.H. Chamseddine, G. Felder, J. Fröhlich, "Gravity in Non-Commutative Geometry", *Comm. Math. Phys.* **155** (1993) 205.

[36] A.H. Chamseddine, J. Fröhlich, O. Grandjean, "The gravitational sector in the Connes-Lott formulation of the standard model", *J. Math. Phys.* **36** (1995) 6255.

[37] Vyjayanthi Chari, Andrew Pressley, *A Guide to Quantum Groups*, Cambridge University Press, 1994.

[38] S. Cho, "Quantum Mechanics on the h-deformed Quantum Plane", *J. Phys. A: Math. Gen.* **32** (1999) 2091.

[39] S. Cho, R. Hinterding, J. Madore, H. Steinacker, "Finite Field Theory on Noncommutative Geometries", Munich Preprint, MPI-PhT/99-12, hep-th/9903239.

[40] S. Cho, K.S. Park, "Linear Connections on Graphs", *J. Math. Phys.* **38** (1997) 5889.

[41] S. Cho, J. Madore, K.S. Park, "Noncommutative Geometry of the h-deformed Quantum Plane", *J. Phys. A: Math. Gen.* **31** (1998) 2639.

[42] A. Connes, "Non-Commutative Differential Geometry", *Publications of the Inst. des Hautes Etudes Scientifiques* **62** (1986) 257.

[43] A. Connes, "The Action Functional in Non-Commutative Geometry", *Comm. Math. Phys.* **117** (1988) 673.

[44] A. Connes, *Essay on Physics and Non-commutative Geometry*, The Interface of Mathematics and Particle Physics, Oxford, September 1988, Clarendon Press, Oxford, 1990.

[45] A. Connes, *Noncommutative Geometry*, Academic Press, 1994.

[46] A. Connes, "Noncommutative geometry and reality", *J. Math. Phys.* **36** (1995) 6194.

[47] A. Connes, "Gravity coupled with matter and the foundation of non-commutative geometry", *Comm. Math. Phys.* **192** (1996) 155.

[48] A. Connes, J. Lott, "Particle Models and Noncommutative Geometry", Recent Advances in Field Theory, *Nucl. Phys. (Proc. Suppl.)* **B18** (1990) 29.

[49] A. Connes, J. Lott, *The metric aspect of non-commutative geometry*, Proceedings of the Cargèse Summer School, July 1991, Plenum Press, New York, 1992.

[50] A. Connes, M.A. Rieffel, "Yang-Mills for non-commutative two-tori", *Contemp. Math.* **62** (1987) 237.

[51] A. Connes, Michael R. Douglas, Albert Schwarz, "Noncommutative Geometry and Matrix Theory: Compactification on Tori", *J. High Energy Phys.* **02** (1998) 003.

[52] R. Coquereaux, "Noncommutative geometry and theoretical physics", *J. Geom. Phys.* **6** (1989) 425.

[53] R. Coquereaux, G. Esposito-Farèse, G. Vaillant, "Higgs Fields as Yang-Mills Fields and Discrete Symmetries", *Nucl. Phys.* **B353** (1991) 689.

[54] L. Crane, L. Smolin, "Space-Time Foam As The Universal Regulator", *Gen. Rel. Grav.* **17** (1985) 1209.

[55] A. Cuntz, D. Quillen, "Algebra extensions and nonsingularity", *J. Amer. Math. Soc.* **8** (1995) 251.

[56] L. Dabrowski, P.M. Hajac, G. Landi, P. Siniscalco, "Metrics and pairs of left and right connections on bimodules", *J. Math. Phys.* **37** (1996) 4635.

[57] S. Deser, "General Relativity and the Divergence Problem in Quantum Field Theory", *Rev. Mod. Phys.* **29** (1957) 417.

[58] B. de Wit, J. Hoppe, H. Nicolai, "On the Quantum Mechanics of Supermembranes", *Nucl. Phys.* **B305** (1988) 545.

[59] A. Dimakis, J. Madore, "Differential Calculi and Linear Connections", *J. Math. Phys.* **37** (1996) 4647.

[60] P.A.M. Dirac, "On Quantum Algebras", *Proc. Camb. Phil. Soc.* **23** (1926) 412.

[61] H.-D. Doebner, J.-D. Hennig (Eds.), *Quantum Groups*, Lect. Notes in Phys. No. 370, Springer-Verlag, Heidelberg, 1990,

[62] S. Doplicher, K. Fredenhagen, J.E. Roberts, "The Quantum Structure of Spacetime at the Planck Scale and Quantum Fields", *Comm. Math. Phys.* **172** (1995) 187.

[63] M. Douglas, "Two Lectures on *D*-Geometry and Noncommutative Geometry", ICTP Spring School Lectures, hep-th/9901146.

[64] M. Douglas, C. Hull, "*D*-Branes and the Noncommutative Torus", *J. High Energy Phys.* **02** (1998) 008.

[65] V.G. Drinfeld, "Quantum Groups", *I.C.M. Proceedings, Berkeley* (1986) 798; "*idem*", *Sov. Math. Dokl.* **36** (1988) 212.

[66] M. Dubois-Violette, "Dérivations et calcul différentiel non-commutatif", *C. R. Acad. Sci. Paris* **307** (1988) 403.

[67] M. Dubois-Violette, R. Kerner, J. Madore, "Gauge bosons in a noncommutative geometry", *Phys. Lett.* **B217** (1989) 485.

[68] M. Dubois-Violette, J. Madore, T. Masson, J. Mourad, "Linear Connections on the Quantum Plane", *Lett. Math. Phys.* **35** (1995) 351.

[69] M. Dubois-Violette, J. Madore, T. Masson, J. Mourad, "On Riemann Curvature in Noncommutative Geometry", *J. Math. Phys.* **37** (1996) 4089.

[70] A. Einstein, "Riemann-Geometrie und Aufrecherhaltung des Begriffs des Fernparallelismus", *Berliner Ber.* (1928) 219.

[71] H. Ewen, O. Ogievetsky, "Classification of the $GL(3)$ Quantum Matrix Groups", Preprint, q-alg/9412009.

[72] L.D. Faddeev, N.Y. Reshetikhin, L.A. Takhtajan, "Quantization of Lie Groups and Lie Algebras", *Alge. i Analy.* **1** (1989) 178; "*idem*", *Leningrad Math. J.* **1** (1990) 193.

[73] M. Fichtmüller, A. Lorek, J. Wess, "q-Deformed Phase Space and its Lattice Structure", *Z. Physik C - Particles and Fields* **71** (1996) 533.

[74] T. Filk, "Divergencies in a field theory on quantum space", *Phys. Lett.* **376** (1996) 53.

[75] D.R. Finkelstein, *Quantum Relativity : A Synthesis of the Ideas of Einstein and Heisenberg*, Springer-Verlag, Heidelberg, 1996.

[76] G. Fiore, "Quantum Groups $SO_q(N)$, $Sp_q(n)$ have q-Determinant, too", *J. Phys. A: Math. Gen.* **27** (1994) 1.

[77] G. Fiore, "The q-Euclidean Algebra $U_q(e^N)$ and the Corresponding q-Euclidean Lattice", *Int. J. Mod. Phys. A* **11** (1996) 863.

[78] G. Fiore, J. Madore, "Leibniz Rules and Reality Conditions", Preprint 98-13, Dip. Matematica e Applicazioni, Università di Napoli, math/9806071.

[79] G. Fiore, J. Madore, "The Geometry of Quantum Euclidean Spaces", LMU Preprint LMU-TPW 97/23, math/9904027.

[80] J. Fröhlich, T. Kerler, *Quantum Groups, Quantum Categories and Quantum Field Theory*, Lect. Notes in Math. No. 1542, Springer-Verlag, Heidelberg, 1993.

[81] J. Fröhlich, O. Grandjean, A. Recknagel, "Supersymmetric quantum theory, non-commutative geometry and gravitation", Lecture Notes, Les Houches 1995, hep-th/9706132.

[82] R. Gambini, J. Pullin, A. Ashtekar, *Loops, Knots, Gauge Theories and Quantum Gravity*, Cambridge University Press, 1996,

[83] O.J. Ganor, S. Ramgoolam, W. Taylor, "Branes, Fluxes and Duality in M(atrix)-Theory", *Nucl. Phys.* **B492** (1997) 191.

[84] L.J. Garay, "Quantum Gravity and Minimum Length", *Int. J. Mod. Phys. A* **10** (1995) 145.

[85] J.P. Gauntlett, J. Gomis, P.K. Townsend, "BPS Bounds for Worldvolume Branes", *J. High Energy Phys.* **9801** (1998) 003.

[86] Y. Georgelin, J. Madore, T. Masson, J. Mourad, "On the non-commutative Riemannian geometry of $GL_q(n)$", *J. Math. Phys.* **38** (1997) 3263.

[87] B. Greene, S.-T Yau (Eds.), *Mirror Symmetry II*, AMS/IP Studies in Advanced Mathematics, 1997.

[88] H. Grosse, *Models in Mathematical Physics and Quantum Field Theory*, Trieste Notes in Physics, Springer-Verlag, Heidelberg, 1988.

[89] H. Grosse, P. Falkensteiner, "On the relation between the geometric phase and the Schwinger term", *Phys. Lett.* **B253** (1991) 395.

[90] H. Grosse, J. Madore, "A Noncommutative Version of the Schwinger Model", *Phys. Lett.* **B283** (1992) 218.

[91] H. Grosse, P. Prešnajder, "The Construction of Noncommutative Manifolds Using Coherent States", *Lett. Math. Phys.* **28** (1993) 239.

[92] H. Grosse, P. Prešnajder, "The Dirac Operator on the Fuzzy Sphere", *Lett. Math. Phys.* **33** (1995) 171.

[93] H. Grosse, W. Maderner, C. Reitberger, "Schwinger Terms and Cyclic Cohomology for Massive $(1+1)$-dimensional Fermions and Virasoro Algebras", *J. Math. Phys.* **34** (1993) 4469.

[94] H. Grosse, C. Klimčík, P. Prešnajder, *Simple Field Theoretical Models on Noncommutative Manifolds*, Proceedings of the 2nd Bulgarian Workshop on New Trends in Quantum Field Theory, Razlog, August 1995, Heron Press, 1996.

[95] H. Grosse, C. Klimčík, P. Prešnajder, "Topological Nontrivial Field Configurations in Noncommutative Geometry", *Comm. Math. Phys.* **178** (1997) 507.

[96] H. Grosse, C. Klimčík, P. Prešnajder, "On 4D Field Theory in Non-Commutative Geometry", *Comm. Math. Phys.* **180** (1997) 429.

[97] H. Grosse, C. Klimčík, P. Prešnajder, "Field Theory on a Supersymmetric Lattice", *Comm. Math. Phys.* **185** (1997) 155.

[98] H. Grosse, G. Reiter, "The Fuzzy supersphere", *J. Geom. Phys.* **28** (1998) 349.

[99] H. Grosse, G. Reiter, "Graded Differential Geometry of Graded Matrix Algebras", Vienna Preprint UWThPh-6-1999, math-ph/9905018.

[100] P.M. Hajac, "Strong Connections on Quantum Principal Bundles", *Comm. Math. Phys.* **182** (1996) 579.

[101] A. Hebecker, S. Schreckenberg, J. Schwenk, W. Weich, J. Wess, "Representations of a q-deformed Heisenberg algebra", *Z. Physik C - Particles and Fields* **64** (1994) 355.

[102] H. Heckenberger, K. Schmüdgen, "Levi-Civita Connections on the Quantum Groups $SL_q(N)$, $O_q(N)$ and $SP_q(N)$", *Comm. Math. Phys.* **185** (1997) 177.

[103] E.J. Hellund, K. Tanaka, "Quantized Space-Time", *Phys. Rev.* **94** (1954) 192.

[104] Pei-Ming Ho, Yong-Shi Wu, "Noncommutative Geometry and D-branes", *Phys. Lett.* **B398** (1997) 52.

[105] J. Hoppe, "Diffeomorphism groups, Quantization and $SU(\infty)$", *Int. J. Mod. Phys. A* **4** (1989) 5235.

[106] B. Iochum, T. Schücker, "A Left-Right Symmetric Model à la Connes-Lott", *Lett. Math. Phys.* **32** (1994) 153.

[107] C.J. Isham, Abdus Salam, J. Strathdee, "Infinity Suppression Gravity Modified Quantum Electrodynamics", *Phys. Rev.* **D3** (1971) 1805.

[108] A. Jevicki, S. Ramgoolam, "Non commutative gravity from the ADS/CFT correspondence", Brown-Het-1159, **hep-th/9902059**.

[109] M. Jimbo, "A q-difference analogue of $U(\mathfrak{g})$ and the Yang-Baxter equation", *Lett. Math. Phys.* **10** (1985) 63.

[110] M. Kaku, *Generally Covariant Lattices, the Random Calculus, and the Strong Coupling Approach to the Renormalisation of Gravity*, Quantum Field Theory and Quantum Statistics, Vol. 2, I.A. Batalin, C.J. Isham and G.A. Vilkovisky (Eds.), Adam Hilger, Bristol, 1987.

[111] W. Kalau, M. Walze, "Gravity, Noncommutative Geometry and the Wodzicki Residue", *J. Geom. Phys.* **16** (1995) 327.

[112] W. Kalau, N.A. Papadopoulos, J. Plass, J.-M. Warzecha, "Differential Algebras in Non-Commutative Geometry", *J. Geom. Phys.* **16** (1995) 149.

[113] C. Kassel, *Quantum Groups*, Springer-Verlag, Heidelberg, 1995.

[114] D. Kastler, "A detailed account of Alain Connes' version of the standard model in non-commutative geometry", *Rev. Math. Phys.* **5** (1993) 477.

[115] D. Kastler, "The Dirac Operator and Gravitation", *Comm. Math. Phys.* **166** (1995) 633.

[116] D. Kastler, *Noncommutative Geometry and Fundamental Physical Interactions*, Proceedings of the 38. Internationale Universitätswochen für Kern- und Teilchenphysik, 'Geometrie und Quantenphysik', Schladming, Austria, H. Gausterer, H. Grosse, L. Pittner (Eds.), Lect. Notes in Phys., Springer-Verlag, Heidelberg, 1999.

[117] D. Kastler, J. Madore, D. Testard, "Connections of bimodules in noncommutative geometry", *Contemp. Math.* **203** (1997) 159.

[118] D. Kastler, T. Masson, J. Madore, "On finite differential calculi", *Contemp. Math.* **203** (1997) 135.

[119] Y. Katayama, H. Yukawa, "Field Theory of Elementary Domains and Particles I", *Prog. Theor. Phys. (Suppl.)* **41** (1968) 1; "*idem* II", *ibid* **41** (1968) 22.

[120] A. Kehagias, P. Meessen, G. Zoupanos, "Deformed Poincaré Algebra and Field Theory", *Phys. Lett.* **B346** (1995) 262.

[121] A. Kehagias, J. Madore, J. Mourad, G. Zoupanos, "Linear Connections in Extended Space-Time", *J. Math. Phys.* **36** (1995) 5855.

[122] A. Kempf, *On the Structure of Space-Time at the Planck Scale*, Subnuclear Physics: 36th Course: 'From the Planck Length to the Hubble Radius', Erice, World Scientific Publishing, (to appear).

[123] A. Kempf, G. Mangano, "Minimal Length Uncertainty Relation and Ultraviolet Regularization", *Phys. Rev.* **D55** (1997) 7909.

[124] A. Kempf, G. Mangano, R.B. Mann, "Hilbert space representation of the minimal length uncertainty relation", *Phys. Rev.* **D52** (1995) 1108.

[125] A. Klymik, K. Schmüdgen, *Quantum groups and their Representations*, Springer-Verlag, Heidelberg, 1997.

[126] P. Kosinski, J. Lukierski, P. Maslanka, "Local $D = 4$ Field Theory on κ-Deformed Minkowski Space", Preprint FTUV-99-05, hep-th/9902037.

[127] J.L. Koszul, *Lectures on Fibre Bundles and Differential Geometry*, Tata Institute of Fundamental Research, Bombay, 1960.

[128] T. Krajewski, "Classification of Finite Spectral Triples", *J. Geom. Phys.* **28** (1998) 1.

[129] T. Krajewski, R. Wulkenhaar, "Perturbative Quantum Gauge Fields on the Noncommutative Torus", Marseille preprint CPT-99/P.3794, hep-th/9903187.

[130] P.P. Kulish (Ed.), *Quantum Groups*, Lect. Notes in Math. No. 1510, Springer-Verlag, Heidelberg, 1991.

[131] P.P. Kulish, *On recent progress in quantum groups: an introductory review*, Jahrbuch Überblicke Mathematik, 1993, Vieweg Verlag, Braunschweig, Germany, 1993.

[132] P.P. Kulish, E.D. Damaskinsky, "On the q-oscillator and the quantum algebra $su_q(1,1)$", *J. Phys. A: Math. Gen.* **23** (1990) L415.

[133] P.P. Kulish, N.Yu. Reshetikhin, "Quantum Linear Problem for the Sine-Gordon Equation and Higher Representations", *J. Soviet Math.* **23** (1983) 2435.

[134] B.A. Kupershmit, "The quantum group $GL_h(2)$", *J. Phys. A: Math. Gen.* **25** (1992) L1239.

[135] G. Landi, Ai Viet Nguyen, K.C. Wali, "Gravity and electromagnetism in noncommutative geometry", *Phys. Lett.* **B326** (1994) 45.

[136] G, Landi, *An Introduction to Noncommutative Spaces and their Geometries*, Lecture Notes in Physics. New Series M, Monographs No. 51, Springer-Verlag, Heidelberg, 1997.

[137] M. Li, T. Yoneya, "Short Distance Space-Time Structure and Black Holes in String Theory: a Short Review of the Present Status", EFI-98-24, hep-th/9806240.

[138] F. Lizzi, G. Mangano, G. Miele, G. Sparano, "Constraints on Unified Gauge Theories from Noncommutative Geometry", *Mod. Phys. Lett. A* **11** (1996) 2561.

[139] A. Lorek, A. Ruffing, J. Wess, "A q-Deformation of the Harmonic Oscillator", *Z. Physik C - Particles and Fields* **74** (1997) 369.

[140] A. Lorek, W. Weich, J. Wess, "Non-commutative Euclidean and Minkowski Structures", *Z. Physik C - Particles and Fields* **76** (1997) 375.

[141] J. Lukierski, A. Nowicki, H. Ruegg, "New Quantum Poincaré Algebra and κ-deformed Field Theory", *Phys. Lett.* **B293** (1992) 344.

[142] J. Lukierski, H. Ruegg, W. Ruhl, "From κ-Poincaré algebra to κ-Lorentz quasigroup: a Deformation of relativistic Symmetry", *Phys. Lett.* **B313** (1993) 357.

[143] J. Lukierski, P. Minnaert, M. Mozrzymas, "Quantum Deformations of Conformal Algebras Introducing Fundamental Mass Parameters", *Phys. Lett.* **B371** (1996) 215.

[144] A.J. Macfarlane, "On q-analogues of the quantum harmonic oscillator and the quantum group $SU_q(2)$", *J. Phys. A: Math. Gen.* **22** (1989) 4581.

[145] J. Madore, *Non-Commutative Geometry and the Spinning Particle*, Proceedings of the Journées Relativistes, Genève, 1988; *idem*, Proceedings of the XI Warsaw Symposium on Elementary Particle Physics, Kazimierz, May 1988, World Scientific Publishing, 1989.

[146] J. Madore, *Kaluza-Klein Aspects of Noncommutative Geometry*, Proceedings of the XVII International Conference on Differential Geometric Methods in Theoretical Physics, Chester, August 1988, World Scientific Publishing, 1989.

[147] J. Madore, "Modification of Kaluza-Klein Theory", *Phys. Rev.* **D41** (1990) 3709.

[148] J. Madore, "The Fuzzy Sphere", *Class. and Quant. Grav.* **9** (1992) 69.

[149] J. Madore, "On Poisson Structure and Curvature", Proceedings of the XVI Workshop on Geometric Methods in Physics, Bialowieza, July, 1997, *Rep. on Mod. Phys.* **43** (1999) 231.

[150] J. Madore, *Introduction to Non-Commutative Geometry*, Proceedings of the Corfu Summer Institute on Elementary Partticle Physics, September 1998, PRHEP-corfu98/016, 1999.

[151] J. Madore, *An Introduction to Noncommutative Geometry*, Proceedings of the 38. Internationale Universitätswochen für Kern- und Teilchenphysik, 'Geometrie und Quantenphysik', Schladming, Austria, H. Gausterer, H. Grosse, L. Pittner (Eds.), Lect. Notes in Phys., Springer-Verlag, Heidelberg, 1999.

[152] J. Madore, *An Introduction to Noncommutative Differential Geometry and its Physical Applications*, 2nd edition Cambridge University Press, 1999.

[153] J. Madore, J. Mourad, "Algebraic-Kaluza-Klein Cosmology", *Class. and Quant. Grav.* **10** (1993) 2157.

[154] J. Madore, J. Mourad, "On the Origin of Kaluza-Klein Structure", *Phys. Lett.* **B359** (1995) 43.

[155] J. Madore, J. Mourad, "Noncommutative Kaluza-Klein Theory", Lecture given at the 5th Hellenic School and Workshops on Elementary Particle Physics, hep-th/9601169.

[156] J. Madore, J. Mourad, "Quantum Space-Time and Classical Gravity", *J. Math. Phys.* **39** (1998) 423.

[157] J. Madore, L.A. Saeger, "Topology at the Planck Length", *Class. and Quant. Grav.* **15** (1998) 811.

[158] J. Madore, T. Masson, J. Mourad, "Linear Connections on Matrix Geometries", *Class. and Quant. Grav.* **12** (1995) 1429.

[159] J. Madore, J. Mourad, A. Sitarz, "Deformations of Differential Calculi", *Mod. Phys. Lett. A* **12** (1997) 975.

[160] A. Maes, A. Van Daele, "Notes on Compact Quantm Groups", Leuven Preprint, math/9803122.

[161] S. Majid, "Hopf Algebras for physics at the Planck scale", *Class. and Quant. Grav.* **5** (1988) 1587.

[162] S. Majid, "Braided Momentum in the q-Poincaré group", *J. Math. Phys.* **34** (1993) 2045.

[163] S. Majid, *Foundations of Quantum Group Theory*, Cambridge University Press, 1995.

[164] S. Majid, *Quantum Geometry and the Planck Scale*, Group XXI Physical Applications and Mathematical Aspects of Geometry, Groups, and Algebras I, World Scientific Publishing, 1997.

[165] G. Maltsiniotis, "Le Langage des Espaces et des Groupes Quantiques", *Comm. Math. Phys.* **151** (1993) 275.

[166] G. Mangano, "Path Integral Approach to Noncommutative Space-Time", *J. Math. Phys.* **39** (1998) 2584.

[167] Yu.I. Manin, *Quantum Groups and Non-commutative Geometry*, Les Publications du Centre de Recherche Mathématiques, Montréal, 1988.

[168] Yu.I. Manin, "Multiparametric Quantum Deformations of the General Linear Supergroup", *Comm. Math. Phys.* **123** (1989) 163.

[169] C.P. Martín, D. Sánchez-Ruiz, "The One-loop UV-Divergent Structure of $U(1)$-Yang-Mills Theory on Noncommutative $\mathbb{R}^4$", Madrid Preprint FT/UCM-15-99, hep-th/9903077.

[170] H. Moscovici, "Eigenvalue Inequalities and Poincaré Duality in Noncommutative Geometry", *Comm. Math. Phys.* **194** (1997) 619.

[171] J. Mourad, "Linear Connections in Non-Commutative Geometry", *Class. and Quant. Grav.* **12** (1995) 965.

[172] O. Ogievetsky, *Hopf Structures on the Borel subalgebra of $sl(2)$*, Winter School Geometry and Physics, Zidkov, Suppl. Rendiconti cir. Math. Palermo, 185, 1992.

[173] O. Ogievetsky, B. Zumino, "Reality in the Differential Calculus on q-Euclidean Spaces", *Lett. Math. Phys.* **25** (1992) 121.

[174] O. Ogievetsky, W.B. Schmidke, J. Wess, B. Zumino, "q-deformed Poincaré algebra", *Comm. Math. Phys.* **150** (1992) 495.

[175] A. Pais, G.E Uhlenbeck, "On Field Theories with Non-Localized Action", *Phys. Rev.* **79** (1950) 145.

[176] M. Paschke, A. Sitarz, "Discrete spectral triples and their symmetries", *J. Math. Phys.* **39** (1998) 6191.

[177] L. Pittner, *Algebraic Foundations of Non-Commutative Differential Geometry and Quantum Groups*, Springer-Verlag, Heidelberg, 1998.

[178] P. Podleś, "Quantum Spheres", *Lett. Math. Phys.* **14** (1987) 193.

[179] P. Podleś, "Solutions of the Klein-Gordon and Dirac Equations on Quantum Minkowski Spaces", *Comm. Math. Phys.* **181** (1996) 569.

[180] P. Podleś, S.L. Woronowicz, "Quantum Deformation of Lorentz Group", *Comm. Math. Phys.* **130** (1990) 381.

[181] P. Podleś, S.L. Woronowicz, "On the Classification of Quantum Poincaré Groups", *Comm. Math. Phys.* **178** (1996) 61.

[182] I. Pris, T. Schücker, "Noncommutative geometry beyond the standard model", *J. Math. Phys.* **38** (1997) 2255.

[183] W. Pusz, S.L. Woronowicz, "Twisted Second Quantization", *Rev. Mod. Phys.* **27** (1989) 231.

[184] M.A. Rieffel, "Non-commutative Tori - A case study of Non-commutative Differentiable Manifolds", *Contemp. Math.* **105** (1980) 191.

[185] A.D. Sakharov, "Vacuum quantum fluctuations in curved space and the theory of gravitation", *Doklady Akad. Nauk. S.S.S.R.* **177** (1967) 70.

[186] A.D. Sakharov, "Spectral Density of Eigenvalues of the Wave Equation and Vacuum Polarization", *Teor. i Mat. Fiz.* **23** (1975) 178.

[187] M. Schlieker, W. Weich, R. Weixler, "Inhomogeneous Quantum Groups", *Z. Physik C - Particles and Fields* **53** (1992) 79.

[188] W.B. Schmidke, J. Wess, B. Zumino, "A q-deformed Lorentz algebra", *Z. Physik C - Particles and Fields* **52** (1991) 471.

[189] K. Schmüdgen, "Operator representations of a q-deformed Heisenberg algebra", Leipzig Preprint, math/9805131.

[190] K. Schmüdgen, "On the construction of covariant differential calculi on quantum homogeneous spaces", *J. Geom. Phys.* **30** (1999) 23-47.

[191] K. Schmüdgen, "Commutator Representations of Differential Calculi on Quantum Group $SU_q(2)$", *J. Geom. Phys.* (1999)

[192] V. Schomerus, "D-branes and Deformation Quantization", DESY Preprint 99-039, hep-th/9903205.

[193] E. Schrödinger, "Über die Kräftefreie Bewegung in der Relativistischen Quantenmechanik", *Sitzungber. Preuss. Akad. Wiss. Phys.-Math. Kl.* **24** (1930) 418; *Gesammelte Abhandlungen*, Friedr. Vieweg & Sohn Verlag, Braunschweig/Wiesbaden, 1984.

[194] T. Schücker, *Geometries and Forces*, Proceedings of the European Mathematical Society Summer School on Noncommutative Geometry and Applications, Monsaraz and Lisbon, Portugal, September, 1997.

[195] P. Schupp, P. Watts, B. Zumino, "The 2-Dimensional Quantum Euclidean Algebra", *Lett. Math. Phys.* **24** (1992) 141.

[196] J. Schwenk, J. Wess, "A q-deformed Quantm Mechanical Toy Model", *Phys. Lett.* **B291** (1992) 273.

[197] J. Schwinger, "Unitary Operator Bases", *Proc. Nat. Acad. Sci. (USA)* **46** (1960) 570.

[198] I. Segal, "Quantized differential forms", *Topology* **7** (1968) 147.

[199] E.K. Sklyanin, "Some algebraic structures connected with the Yang-Baxter equation", *Funct. Anal. Appl.* **16** (1982) 263.

[200] J.J. Sylvester, "Lectures on the Principles of Universal Algebra", *Amer. J. Math.* **6** (1884) 271.

[201] H.S. Snyder, "Quantized Space-Time", *Phys. Rev.* **71** (1947) 38.

[202] H.S. Snyder, "The Electromagnetic Field in Quantized Space-Time", *Phys. Rev.* **72** (1947) 68.

[203] H. Steinacker, "Finite Dimensional Unitary representations of Quantum Anti-de Sitter Groups at Roots of Unity", *Comm. Math. Phys.* **192** (1998) 687.

[204] László Szabó, *Geometry of Quantum Space-Time*, Proceedings of the XI Warsaw Symposium on Elementary Particle Physics, Kazimierz, May 1988, World Scientific Publishing, 1989.

[205] Y. Takano, *Some Topics and Trials on the Theory of Non-Local Fields - Theory of Fields in Finsler Spaces and Fluctuation of Space-Time*, International School of Gravitation, Erice: 16th Course: 'Classical and Quantum Non-Locality', World Scientific Publishing, (to appear).

[206] W. Taylor, "Lectures on D-branes, Gauge Theory and M(atrices)", 2nd Trieste Conference on Duality in String Theory, June 1997, hep-th/9801182.

[207] G. 't Hooft, "Quantization of point particles in (2+1)-dimensional gravity and spacetime discreteness", *Class. and Quant. Grav.* **13** (1996) 1023.

[208] J.C. Várilly, *Introduction to Noncommutative Geometry*, Proceedings of the European Mathematical Society Summer School on Noncommutative Geometry and Applications, Monsaraz and Lisbon, Portugal, September, 1997.

[209] J.C. Várilly, J.M. Gracia-Bondía, "Connes' noncommutative differential geometry and the Standard Model", *J. Geom. Phys.* **12** (1993) 223.

[210] J.C. Várilly, J.M. Gracia-Bondía, "On the ultraviolet behaviour of quantum fields over noncommutative manifolds", Costa-Rica Preprint, hep-th/9804001.

[211] J. Wess, B. Zumino, "Covariant Differential Calculus on the Quantum Hyperplane", *Nucl. Phys. (Proc. Suppl.)* **18B** (1990) 302.

[212] H. Weyl, *The Theory of Groups and Quantum Mechanics*, Dover, New York, 1950.

[213] S.L. Woronowicz, *Pseudospaces, pseudogroups and Pontriagin duality*, Proceedings of the International conference on Mathematics and Physics, Lausanne, Lect. Notes in Phys. No. 116, Springer-Verlag, Heidelberg, 1979.

[214] S.L. Woronowicz, "Twisted $SU(2)$ Group, an example of a Non-Commutative Differential Calculus", *Publ. RIMS, Kyoto Univ.* **23** (1987) 117.

[215] S.L. Woronowicz, "Compact Matrix Pseudogroups", *Comm. Math. Phys.* **111** (1987) 613.

[216] S.L. Woronowicz, "Unbounded Elements Affiliated with C^*-algebras and Non-compact Quantum Groups", *Comm. Math. Phys.* **136** (1991) 399.

[217] C.N. Yang, "On Quantized Space-Time", *Phys. Rev.* **72** (1947) 874.

[218] H. Yukawa, "On the radius of the Elementary Particle", *Phys. Rev.* **76** (1949) 300.

[219] G. Zampieri, A. D'Agnolo (Eds.), *D-Modules, Representation Theory, and Quantum Groups*, Lect. Notes in Math. No. 1565, Springer-Verlag, Heidelberg, 1993.

Some Topics and Trials on the Theory of Non-Local Fields
—Theory of Fields in Finsler Spaces and Fluctuation of Space-Time—

Yoshiro TAKANO

Department of Physics, Yokohama National University

79-5 Tokiwadai, Hodogaya-ku, Yokohama 240-8501, Japan

Abstract

The introduction of non-locality into elementary particles is to solve two fundamental problems, that is, the divergence difficulty and the unified description of elementary particles. Yukawa's theory is referred to as a typical theory of non-local fields.

As a geometrization of Yukawa's theory of non-local fields, the theory of fields in Finsler spaces is developed. Finsler spaces are spaces whose metric tensor $g_{\mu\nu}$ depends on directional variables y^λ as well as coordinates x^λ. In Finsler spaces, any geometrical quantities depend on the directional variables as well as the coordinates, and, therefore, the internal freedom is introduced in the fields of elementary particles and the effects equivalent to those of the non-locality can be expected. Besides the microscopic world, in the macroscopic world, the geometrical structure of space-time is modified by the directional dependence of metric and the cosmological observable effects originating from the anisotropy of spaces and the torsion is expected to exist.

The removal of divergence is discussed in connection with the causality. Spaces of the special theory of relativity do not have a capacity to accept the extended particles and the metric fundamental form should be modified so as to change micro-causality, leaving macro-causality intact. The indefinite metric in Hilbert space presents itself as the result of the probability distribution of the metric fundamental form, the fluctuation of space-time. The variety of elementary particles is connected with the variety of distribution.

Introduction——Yukawa's theory of non-local fields[1]——

As to the theory of non-local fields, though it seems to be of great promise, we are still, regrettably, groping for our way. We would like, therefore, to take this opportunity to present some topics and trials on the theory of non-local fields, which so far we have been concerned with, for further discussions.

It should be convenient for our study to start from Yukawa's theory, a typical theory of non-local fields. Since Yukawa proposed his theory of non-local fields, it has already passed half a century. In his original theory, the equations of non-local scalar field $U(X_\mu, r_\mu)$, for example, are assumed as follows:

$$\begin{cases} \left(\dfrac{\partial^2}{\partial X_\mu \partial X^\mu} + \kappa^2\right) U(X_\mu, r_\mu) = 0, & (1) \\ (r_\mu r^\mu + \lambda^2)\, U(X_\mu, r_\mu) = 0, & (2) \\ r^\mu \dfrac{\partial U(X_\mu, r_\mu)}{\partial X^\mu} = 0, & (3) \end{cases}$$

where X_μ could be identified with the conventional space-time coordinates of the elementary particle and r_μ could be interpreted as the internal coordinates and $\kappa = mc/\hbar$ is the mass, and, as is easily shown, the field $U(X_\mu, r_\mu)$ differs from zero only on the spherical surface of radius λ in the rest system of coordinates.

The introduction of non-locality into elementary particles is, needless to say, to solve the two fundamental problemes, that is, the divergence difficulty and the unified description of elementary particles. In order to increase the degrees of internal freedom, afterward, Yukawa has generalized his theory so that field quantities depend on the internal vector r_μ with variable magnitude, but not fixed magnitude λ, and, further, depend on triple, not single, internal vectors.

Now, we would like to discuss on the geometrization of Yukawa's theory of non-local fields by Finsler geometry, at first, and, secondly, on the removal of divergence in connection with the causality.

Part I. Theory of Fields in Finsler Spaces[2,3,4,5]

§1. Finsler geometry[6]

Roughly speaking, Finsler spaces are spaces whose metric tensor $g_{\mu\nu}$ depends on directional variables y^λ as well as coordinates x^λ. The directional variables y^λ transform like components of a contravariant vector under a coordinate transformation: $x'^\lambda = x'^\lambda(x^\lambda)$. Finsler geometry is also formulated so as to depend only on the direction of vector y^λ, but not its magnitude.

Accordingly, with respect to the infinitesimal changes of the directional variables and the coordinates, respectively, two kinds of affine connections $C_\lambda{}^\mu{}_\nu$ and $\Gamma^*{}_\lambda{}^\mu{}_\nu$ are defined, and further, three kinds of curvature tensors $S_\kappa{}^\lambda{}_{\mu\nu}$, $P_\kappa{}^\lambda{}_{\mu\nu}$ and $R_\kappa{}^\lambda{}_{\mu\nu}$ are defined, in the analogous way as Riemann geometry.

The absolute differential and the covariant derivatives of a vector field $V^\mu(x, y)$, which is assumed to be homogeneous of degree zero in y^λ, are defined as follows:

$$D V^\mu = V^\mu{}_{|\lambda} dx^\lambda + V^\mu{}_{\|\lambda} D l^\lambda, \tag{4}$$

$$\begin{cases} V^\mu{}_{|\lambda} = \dfrac{\partial^* V^\mu}{\partial x^\lambda} + V^\kappa \Gamma^*{}_\kappa{}^\mu{}_\lambda, & (5) \\ V^\mu{}_{\|\lambda} = F \dfrac{\partial V^\mu}{\partial y^\lambda} + V^\kappa F C_\kappa{}^\mu{}_\lambda, & (6) \end{cases}$$

where, $\partial^*/\partial x^\lambda$ is a vector differential operator,

$$\frac{\partial^*}{\partial x^\lambda} = \frac{\partial}{\partial x^\lambda} - N^\kappa{}_\lambda \frac{\partial}{\partial y^\kappa}, \tag{7}$$

and N^κ_λ is called non-linear connection and is given as below:

$$N^\kappa_\lambda = \Gamma^{*\kappa}_{\iota\ \lambda} y^\iota. \tag{8}$$

The metric fundamental function,

$$F^2 = e\, g_{\mu\nu} y^\mu y^\nu, \tag{9}$$

where e is +1 or -1 according as y^λ is time-like or space-like, is multiplied to make any geometrical quantities to be homogeneous of degree zero in y^λ. For example, $l^\lambda = y^\lambda / F$.

The metric tensor is, inversely, given as follows:

$$g_{\mu\nu} = \frac{1}{2} \frac{\partial^2 (e F^2)}{\partial y^\mu \partial y^\nu}. \tag{10}$$

The absolute differential and the covariant derivatives of the metric tensor vanish identically,

$$D g_{\mu\nu} = 0, \tag{11}$$

$$\begin{cases} g_{\mu\nu|\lambda} = 0, & (12) \\ g_{\mu\nu\|\lambda} = 0, & (13) \end{cases}$$

that is, such spaces as above are metrical.

The anisotropy of space is represented by the indicatrix, which is defined by the equation,

$$F(x, y) = 1, \quad x : \text{fixed}, \tag{14}$$

and is a generalization of unit sphere.

§2. Fields of elementary particles

(1) Scalar field and internal freedom

In Finsler spaces, any geometrical quantities depend on directional variables as well as coordinates, and, therefore, the internal freedom is introduced in the fields of elementary particles and the effects equivalent to those of the non-locality can be expected, and even the massless particles can also be treated as the non-local fields, while, being based on the unlimitedly divisible structure of space-time, differential operation can be applied.

For example, in Finsler spaces, the equation of scalar field $\varphi(x, y)$, which is homogeneous of degree zero in y^λ, may be assumed as follows:

$$\left[\Box_{xx} + f\left(F\frac{\partial}{\partial y^\mu}, l^\mu, g_{\mu\nu}, \lambda\right)\right]\varphi(x, y) = 0, \tag{15}$$

where $\Box_{xx} = 1/\sqrt{-g}\,\partial^*/\partial x^\mu(\sqrt{-g}\,g^{\mu\nu}\,\partial^*/\partial x^\nu)$ is the generalized d'Alembertian and the operator f is a certain invariant function of $F\,\partial/\partial y^\mu$, l^μ, $g_{\mu\nu}$ and a constant λ with the dimension of length. This equation is nothing but the generalized equation of non-local scalar field.

When Finsler spaces are reduced to Minkowski spaces and further to Lorentz spaces, spaces of the special theory of relativity, the equation (15) may be separated to the usual Klein-Gordon equation and the eigenvalue equation concerning the internal freedom:

$$\begin{cases}(\Box + \kappa^2)\,\varphi^{(e)}(x) = 0, & (16)\\ \left[f\left(F\frac{\partial}{\partial y^\mu}, l^\mu, \lambda\right) - \kappa^2\right]\varphi^{(i)}(y) = 0, & (17)\end{cases}$$

$$\varphi(x, y) = \varphi^{(e)}(x)\,\varphi^{(i)}(y), \tag{18}$$

where κ is a separation constant, interpreted as the mass, and $\varphi^{(e)}$ and $\varphi^{(i)}$ are the external and the internal states, respectively.

It is also expected that the singularity of propagators might be removed by virtue of the average over directional variables.

(2) Spinor field and associated gauge fields

Spinor field $\psi(x, y)$ is naturally assumed to be homogeneous of degree zero in y^λ, and its absolute differential and covariant derivatives can be defined as follows:

$$D\psi = \psi_{|\lambda} dx^\lambda + \psi_{\|\lambda} Dl^\lambda, \tag{19}$$

$$\begin{cases} \psi_{|\lambda} = \dfrac{\partial^* \psi}{\partial x^\lambda} - \Gamma^*_\lambda \psi, & (20) \\ \psi_{\|\lambda} = F \dfrac{\partial \psi}{\partial y^\lambda} - F C_\lambda \psi, & (21) \end{cases}$$

where Γ^*_λ and C_λ are spin affine connections and the latter is assumed to satisfy the condition,

$$C_\lambda y^\lambda = 0. \tag{22}$$

The equation of spinor field in Finsler spaces may be assumed, for instance, as follows:

$$\left[i\gamma^\mu \left(\frac{\partial^*}{\partial x^\mu} - \Gamma^*_\mu \right) + f\left(F\frac{\partial}{\partial y^\mu} - FC_\mu, \gamma^\mu, l^\mu, g_{\mu\nu}, \lambda \right) \right] \psi(x,y) = 0, \tag{23}$$

where the generalized Dirac Matrices γ^μ depend on the coordinates and the directional variables and λ is a constant with the dimension of length.

The spin affine connections are determined by the requirement that the absolute differential and the covariant derivatives of the generalized Dirac matrices γ^μ vanish identically:

$$D\gamma^\mu = \gamma^\mu{}_{|\lambda} dx^\lambda + \gamma^\mu{}_{\|\lambda} Dl^\lambda = 0, \tag{24}$$

$$\begin{cases} \gamma^\mu{}_{|\lambda} = \dfrac{\partial^* \gamma^\mu}{\partial x^\lambda} + \gamma^\nu \Gamma^{*\mu}_\nu{}_\lambda - \Gamma^*_\lambda \gamma^\mu + \gamma^\mu \Gamma^*_\lambda = 0, & (25) \\ \gamma^\mu{}_{\|\lambda} = F\dfrac{\partial \gamma^\mu}{\partial y^\lambda} + \gamma^\nu F C_\nu{}^\mu{}_\lambda - FC_\lambda \gamma^\mu + \gamma^\mu F C_\lambda = 0, & (26) \end{cases}$$

where $\Gamma^{*\mu}_\nu{}_\lambda$ and $C_\nu{}^\mu{}_\lambda$ are the affine connections. The matrices γ^μ are related to the metric tensor $g_{\mu\nu}$ as follows:

$$\{\gamma^\mu, \gamma^\nu\} = \gamma^\mu\gamma^\nu + \gamma^\nu\gamma^\mu = 2g^{\mu\nu}. \tag{27}$$

The spin affine connections Γ^{*}_{λ} and C_{λ}, however, are not determined unambiguously and these arbitrarinesses can be referred to the gauge transformations, in the same way as the case of Riemann spaces. The arbitrary terms of Γ^{*}_{λ} and C_{λ} are interpreted as the vector gauge fields.

This situation might be imaged in such a way that the indicatrix is an elastic membrane bounding the elementary particle and a gauge transformation of spinor field generates a strain on this membrane, being accompanied by a gauge field as a stress.

It must also be noted that such gauge fields as above can be regarded as either Abelian or non-Abelian. One of these gauge fields may be identified to electromagnetic field and the other to Yang-Mills field.

§3. Gravitational field

(1) Equation of gravitational field --- Directional dependence

Besides the microscopic world, in the macroscopic world, the geometrical structure of space-time will be modified by the directional dependence of metric and the cosmological observable effects originating from the directional dependence will be expected to exist. Special attention should be paid to the anisotropy of spaces and the torsion.

Regarding the y-dependence of the metric tensor $g_{\mu\nu}$, in analogy with its x-dependence, we propose the following equation:

$$\hat{S}_{\mu\nu} - \frac{1}{2} g_{\mu\nu} \hat{S} = -\kappa^{(i)} T^{(i)}_{\mu\nu}, \tag{28}$$

where $\kappa^{(i)}$ and $T^{(i)}_{\mu\nu}$ might be termed the internal gravitational constant and the internal energy-momentum tensor.

The first curvature tensor $\hat{S}_{\nu}{}^{\mu}{}_{\lambda\kappa}$ is given as follows:

$$\hat{S}_{\nu}{}^{\mu}{}_{\lambda\kappa} = F^{-2} S_{\nu}{}^{\mu}{}_{\lambda\kappa} = C_{\alpha}{}^{\mu}{}_{\lambda} C_{\nu}{}^{\alpha}{}_{\kappa} - C_{\alpha}{}^{\mu}{}_{\kappa} C_{\nu}{}^{\alpha}{}_{\lambda}, \tag{29}$$

$$C_{\nu\mu\lambda} = \frac{1}{2}\frac{\partial g_{\nu\mu}}{\partial y^\lambda} = \frac{1}{4}\frac{\partial^3(eF^2)}{\partial y^\nu \partial y^\mu \partial y^\lambda}. \tag{30}$$

The internal energy-momentum tensor $T^{(i)}_{\mu\nu}$ is symmetric and is assumed that its divergence with respect to y vanishes:

$$T^{(i)\mu\nu}{}_{\|\mu} = 0. \tag{31}$$

In other words, the conservation of the internal energy-momentum with respect to y-dependence is assumed. Since the left-hand side of eq. (28) is shown to be divergence-free by means of Bianchi identity, the above condition (31) is consistent with eq. (28).

On the indicatrix, eq. (28) is reduced to Einstein's equation in three-dimensional Riemann spaces, and the internal cosmological term automatically introduced in this equation,

$$R^{(i)}_{ij} - \frac{1}{2} g_{ij} R^{(i)} + g_{ij} = -\kappa^{(i)} T^{(i)}_{ij}, \tag{32}$$

since the first curvature tensor $S_{\nu\mu\lambda\kappa}$ in Finsler spaces is connected with the curvature tensor $R^{(i)}_{hkji}$, resulted from the induced (Riemann) metric g_{ij} on the indicatrix, as follows:

$$R^{(i)}_{hkji} = S_{hkji} + (g_{hj} g_{ki} - g_{hi} g_{kj}). \tag{33}$$

For the vanishing internal energy-momentum,

$$T^{(i)}_{\mu\nu} = 0, \tag{34}$$

eq. (28) is reduced to

$$\hat{S}_{\mu\nu} = 0. \tag{35}$$

In four-dimensional Finsler spaces, eq. (35) is proved to lead necessarily

$$\hat{S}_\nu{}^\mu{}_{\lambda\kappa} = 0, \tag{36}$$

whether the metric is definite or indefinite.

If Brickell's theorem is supposed to hold also in the case of indefinite metric, our physical space-time will be pictured in such a way that it is Finslerian in the region with internal source and is Riemannian in the empty region.

It seems quite natural to suppose that the internal energy-momentum is given by the fields of elementary particles. Then, the elementary particles inspire space-time with the directional dependence and, conversely, in Finsler spaces the elementary particles are inspired with the internal freedom.

(2) Equation of gravitational field

With regard to the x-dependence of the metric tensor $g_{\mu\nu}$, it is natural to assume, as in Riemann spaces, the following equation:

$$R_{\mu\nu} - \frac{1}{2} g_{\mu\nu} R = \kappa T_{\mu\nu} , \tag{37}$$

$$\begin{cases} R_{\underline{\mu\nu}} - \frac{1}{2} g_{\mu\nu} R = \kappa T_{\underline{\mu\nu}} , & (38) \\ R_{\underset{\smile}{\mu\nu}} = \kappa T_{\underset{\smile}{\mu\nu}} . & (39) \end{cases}$$

The third curvature tensor $R_\nu{}^\mu{}_{\lambda\kappa}$ in Finsler spaces is given as follows:

$$R_\nu{}^\mu{}_{\lambda\kappa} = \frac{\partial^* \Gamma^*_\nu{}^\mu{}_\lambda}{\partial x^\kappa} - \frac{\partial^* \Gamma^*_\nu{}^\mu{}_\kappa}{\partial x^\lambda} + \Gamma^*_\iota{}^\mu{}_\kappa \Gamma^*_\nu{}^\iota{}_\lambda - \Gamma^*_\iota{}^\mu{}_\lambda \Gamma^*_\nu{}^\iota{}_\kappa + C_\nu{}^\mu{}_\alpha R_\lambda{}^\alpha{}_\kappa , \tag{40}$$

$$\Gamma^*_{\nu\mu\lambda} = \Gamma_{\nu\mu\lambda} - C_{\nu\mu\kappa} \Gamma_\iota{}^\kappa{}_\lambda y^\iota = \frac{1}{2} \left(\frac{\partial^* g_{\nu\mu}}{\partial x^\lambda} + \frac{\partial^* g_{\mu\lambda}}{\partial x^\nu} - \frac{\partial^* g_{\lambda\nu}}{\partial x^\mu} \right). \tag{41}$$

It consists of three parts, that is, the first part originating from the torsion, the second originating in the y-dependence of the metric tensor $g_{\mu\nu}$ and the connection coefficient $\Gamma^*_\nu{}^\mu{}_\lambda$, and the third which has the same form as one of Riemann spaces, but depends still on the directional variables, different from Riemann spaces. The anti-symmetric part of the contracted third curvature tensor $R_{\underset{\vee}{\mu\nu}}$ is given as below:

$$R_{\underset{\vee}{\mu\nu}} = \frac{1}{2}\left(C_{\mu}{}^{\lambda}{}_{\alpha} R_{\nu}{}^{\alpha}{}_{\lambda} + C_{\nu}{}^{\lambda}{}_{\alpha} R_{\lambda}{}^{\alpha}{}_{\mu} + C_{\lambda}{}^{\lambda}{}_{\alpha} R_{\mu}{}^{\alpha}{}_{\nu} \right), \tag{42}$$

where

$$R_{\lambda}{}^{\alpha}{}_{\kappa} = R_{\beta}{}^{\alpha}{}_{\lambda\kappa}\, y^{\beta} \tag{43}$$

is one of torsion tensors. It must be added that the tensor $C_{\kappa}{}^{\lambda}{}_{\alpha}$ plays a part of torsion, besides a part of affine connection coefficient.

Contracting Bianchi identity, we obtain

$$\left(R^{\mu\nu} - \frac{1}{2} g^{\mu\nu} R\right)_{|\mu} = U^{\nu}, \tag{44}$$

where

$$U_{\nu} = \frac{1}{2} g^{\iota\lambda} \left(\hat{P}_{\iota}{}^{\kappa}{}_{\lambda\alpha} R_{\kappa}{}^{\alpha}{}_{\nu} + \hat{P}_{\iota}{}^{\kappa}{}_{\mu\alpha} R_{\nu}{}^{\alpha}{}_{\lambda} + \hat{P}_{\iota}{}^{\kappa}{}_{\nu\alpha} R_{\lambda}{}^{\alpha}{}_{\kappa} \right). \tag{45}$$

Hence, we have to assume

$$T^{\mu\nu}{}_{|\mu} = \frac{1}{\kappa} U^{\nu}. \tag{46}$$

The vector U_{ν} acts as a source or a sink of the energy-momentum tensor $T_{\mu\nu}$, or acts as a force density.

Namely, the energy-momentum inflows to or outflows from the internal space and the torsion (dislocation, so to speak) of the external space. Such non-conservation of the external energy-momentum as above is not anything but the reflection of the inhomogeneity of space-time due to the existence of torsion.

As to the second curvature tensor $P_{\nu}{}^{\mu}{}_{\lambda\kappa}$, its geometrical meaning has not yet been clarified and, therefore, the question how the equation is constructed from the second curvature tensor still remains unsettled.

(3) Finsler spaces of constant curvature

In Finsler spaces of constant curvature, the torsion tensor is given as follows:

$$R_{\lambda}{}^{\alpha}{}_{\kappa} = K(y_{\lambda} g^{\alpha}_{\kappa} - y_{\kappa} g^{\alpha}{}_{\lambda}), \tag{47}$$

where K is the scalar curvature and a constant. In these spaces, the antisymmetric part of the contracted third curvature tensor vanishes,

$$R_{[\mu\nu]} = 0, \tag{48}$$

and eq. (44) is reduced to the following:

$$(R^{\mu\nu} - \tfrac{1}{2} g^{\mu\nu} R - \tfrac{1}{2} K \hat{S} y^{\mu} y^{\nu})_{|\mu} = 0. \tag{49}$$

Accordingly, we have to assume as

$$(T^{\mu\nu} - \frac{\kappa^{(i)}}{2\kappa} K T^{(i)\lambda}{}_{\lambda} y^{\mu} y^{\nu})_{|\mu} = 0. \tag{50}$$

It means that the total energy-momentum is conserved, if we regard a combination of the internal and the external energy-momentum tensors: $T_{\mu\nu} - \kappa^{(i)}/2\kappa \cdot K T^{(i)\lambda}{}_{\lambda} y_{\mu} y_{\nu}$ as the total energy-momentum tensor: $T_{\mu\nu}^{total}$. Consequently, the external and the internal energy-momentum are connected with each other, through the medium of torsion.

§4. Generalization of Finsler spaces

To increase the degrees of internal freedom, it will be necessary to generalize Finsler spaces in such a way that geometrical quantities depend not only upon the direction of the vector y^{λ} but also upon its magnitude, and, further, depend upon plural, not single, directional vectors. These generalizations can be performed.

It is noted that our theories of fields in Cartan's Finsler spaces, in the generalized Finsler spaces, and further generalized

Finsler spaces, respectively, correspond to the three stages of Yukawa's non-local field theory.

§5. Spaces whose metric tensor depends on spinor variables[7)]

The differential geometry of spaces whose metric tensor depends on spinor variables ξ and $\bar{\xi}$ as well as coordinates x^λ can be formulated in analogy with the differential geometry of Finsler spaces.

The connection coefficients $\Gamma^{*}{}_{\nu}{}^{\mu}{}_{\lambda}$, $C_{\nu}{}^{\mu}$ and $\bar{C}_{\nu}{}^{\mu}$ and the spin connection coefficients Γ^{*}_{λ}, C and $\tilde{C}$ are determined in such a way that the absolute differentials and the covariant derivatives of the metric tensor $g_{\mu\nu}$ and the generalized Dirac matrices γ^λ vanish identically.

The spin connection coefficients Γ^{*}_{λ}, C and $\tilde{C}$, however, are not determined unambiguously and these arbitrarinesses can be referred to gauge transformations, in the same manner as in Riemann and in Finsler spaces. Although the arbitrary term of Γ^{*}_{λ} is interpreted as a vector gauge field, the arbitrary terms of C and $\tilde{C}$ are interpreted as spinor gauge fields, which may intermediate between fermions and bosons.

It must also be noticed that two kinds of local gauge transformations are significant. One of them corresponds to ordinary local phase transformations of spinor fields, in which coordinate variables x^λ, ξ and $\bar{\xi}$ are unchanged. In the other, coordinate spinor variables ξ and $\bar{\xi}$ themselves also undergo local phase transformations and the connection fields are transformed so that the covariant derivatives should be covariant.

Such spaces as above may be regarded as spinor bundles in the same way that Finsler spaces may be regarded as tangent bundles.

Needless to say, the differential geometry of spaces whose metric tensor depends on vector and spinor variables can also be formulated.

Part II. Fluctuation of Space-Time

16. Removal of divergence in connection with causality

As to the divergence difficulty, it is closely connected with the causality.

If the commutation relation of non-local scalar field is assumed as Yukawa's original formalism,

$$[u(k_\mu, l_\mu), u^*(k'_\mu, l'_\mu)] = -\frac{k_4}{|k_4|}\delta(k_\mu k^\mu - \kappa^2)\,\delta(l_\mu l^\mu + \lambda^2)\,\delta(k_\mu l^\mu)\,\delta(k_\mu - k'_\mu)\,\delta(l_\mu - l'_\mu), \qquad (51)$$

where $u(k_\mu, l_\mu)$ is the Fourier transform of $U(X_\mu, r_\mu)$,

$$U(X_\mu, r_\mu) = \int\cdots\int (dk)^4 (dl)^4\, u(k_\mu, l_\mu)\,\delta(k_\mu k^\mu - \kappa^2)\,\delta(l_\mu l^\mu + \lambda^2)\,\delta(k_\mu l^\mu)\exp(ik_\mu X^\mu)\prod_\mu \delta(r_\mu - l_\mu), \qquad (52)$$

the divergence cannot be removed. Because of $\delta(l_\mu - l'_\mu)$ involved in the commutation relation, the interaction takes place only between the parts of particles of the same internal coordinates with each other, as shown in Fig. 1, and, consequently, an extended particle behaves like an assembly of independent point particles.

Another kind of commutation relation such as

$$[u(k_\mu, l_\mu), u^*(k'_\mu, l'_\mu)] = -\frac{k_4}{|k_4|}\delta(k_\mu k^\mu - \kappa^2)\,\delta(l_\mu l^\mu + \lambda^2)\,\delta(k_\mu l^\mu)\,\delta(k_\mu - k'_\mu)\,\delta(l'_\mu l'^\mu + \lambda^2)\,\delta(k'_\mu l'^\mu) \qquad (53)$$

was assumed by Yennie and Rayski. The interaction is averaged over the parts of extended particles and the smearing yields a cut-off factor and the divergence is removed.[1] In the smearing, however, it must be unavoidable to allow partly the propagation of exceeding the light velocity, as shown in Fig. 2, that is, to violate the causality.

Seems to me, Lorentz spaces do not have a capacity to accept the extended particles. It is well known that the rigid body cannot be allowed to exist in Lorentz spaces. Accordingly, it should be necessary to change micro-causality, leaving macro-causality as it is.

Such a trial as above was proposed by Markov.[1] In his theory, the metric fundamental form is assumed as follows:

$$\tau'^2 = \tau^2 - a^2 = x_\mu^2 - a^2, \quad a: \text{constant}, \tag{54}$$

with the consequence that the light cone is modified to the light hyperboloid, as shown in Fig. 3.

Being substituted τ by τ', the usual invariant D-functions, which connect the field quantities at two points with each other, will be expressed as below:

$$\bar{D}(\tau') = \int \bar{\Delta}(\tau, \kappa)\left[\delta(\kappa^2) - \frac{1}{2}|a|^2 \frac{J_1(|a|\kappa)}{|a|\kappa}\right] d\kappa^2 \tag{55}$$

and the similar expressions for $D^{(1)}$ and D-functions. The above modification of D-functions results in the appearance of the indefinite metric in Hilbert space.

§7. Fluctuation of space-time[10,11,12]

On the other hand, if all elementary particles have the extended structure, a point or a line cannot be observed. The extended structure of elementary particles can never be compatible with the unlimitedly (infinitely) divisible structure of space-time, but with the limitedly (finitely) divisible structure of space-time. And, we have to formulate our physical theory only with observable quantities, excluding unobservable quantities. Our physical space-time, therefore, must be vague, but not sharp.

Now, let us make a postulate that if the elementary particle exists, space-time is distributed, or fluctuated, and the metric fundamental form is given as follows:

$$\tau'^2 = x_\mu'^2 = \tau^2 - a^2 = x_\mu^2 - a^2, \tag{56}$$

a having the probability distribution $p(a)$, which is normalized as

$$\int_{-\infty}^{\infty} p(a)\,da = 1, \tag{57}$$

as it should be, and may also be assumed to be even function of a.

Such a modification of the metric fundamental form as stated above may be understood geometrically as the introduction of the fifth-dimension, which is conditioned by the probability distribution. The fifth-dimension is a probability space as well as a metric space. Namely, our fluctuating space is structurally a five-dimensional probability distribution $p(a)$ of four-dimensional subspces $\tau'^2 = x_r'^2 = x_r^2 - a^2$ corresponding to various values of a.

The distribution, or the fluctuation, of space-time is regarded as a physical concept characterized by a new constant, the so-called universal length. Consequently, any physical quantities in our theory can be defined as a weighted average of the corresponding quantity in each four-dimensional subspace.

In our fluctuating space, therefore, the D-functions will be given as follows:

$$\overline{D}'(\tau) = \int \overline{D}(\tau') p(a) da \tag{58}$$

and the similar expressions for $D^{(1)}$ and D-functions.

If we choose such a distribution as

$$p(a) = \frac{1}{2} \kappa_0 J_1(\kappa_0 |a|), \quad \kappa_0 : \text{constant}, \tag{59}$$

we will obtain the following D-functions,

$$\int_{-\infty}^{\infty} \overline{D}(\tau') \frac{1}{2} \kappa_0 J_1(\kappa_0 |a|) da = \overline{D}(\tau) - \overline{\Delta}(\tau, \kappa_0), \tag{60}$$

$$\int_{-\infty}^{\infty} D^{(1)}(\tau') \frac{1}{2} \kappa_0 J_1(\kappa_0 |a|) da = D^{(1)}(\tau) - \Delta^{(1)}(\tau, \kappa_0), \tag{61}$$

$$\int_{-\infty}^{\infty} D(\tau') \frac{1}{2} \kappa_0 J_1(\kappa_0 |a|) da = D(\tau) - \Delta(\tau, \kappa_0), \tag{62}$$

and the static potential,

$$\int_{-\infty}^{\infty} \frac{e}{\sqrt{r^2 + a^2}} \frac{1}{2} \kappa_0 J_1(\kappa_0 |a|) da = \frac{e}{r}(1 - e^{-\kappa_0 r}). \tag{63}$$

These are nothing but the D-functions and the static potential of Bopp's electrodynamics, which was once interpreted as a mixed field with negative energy.[13]

For another example, the so-called Feynman cut-off,[14] $\frac{1}{\pi}\cdot\kappa_0 \exp(-\kappa_0|\tau|)$, can be derived from the non-negative distribution, $\frac{1}{2}\cdot\kappa_0^2|a|\exp(-\kappa_0|a|)$.

The removal of the singularity of propagators by the fluctuation of space-time is not anything but the introduction of the indefinite metric in Hilbert space or of the negative probability, either of which should be dealt with so as to present itself as the result of some other kind of physical cause.

The reason why the removal of singularity has to appear as the indefinite metric is that the propagation of the extended particle as a whole is forced to express by the propagation of the representative point only. To illustrate this situation, as a simple example, let us expand the static potential $e/\sqrt{r^2+a^2}$, shown in Fig.4, into power-series with respect to the distance r ,

$$\frac{e}{\sqrt{r^2+a^2}} = \frac{e}{r} - \frac{a^2 e}{2r^3} + \cdots, \qquad (64)$$

then we will find a negative charge, or a negative probability of existence of positive charge.

Consequently, the manifestly covariant local theory with indefinite metric is shown to be equivalent to some non-local theory with definite metric.

In our theory, the existence of elementary particles is assumed to produce the distribuition, or the fluctuation, of space-time, and, in turn, the motion of elementary particles must be influenced by the fluctuation of space-time. That is, the existence of matter and the structure of space-time can never be independent of each other.

§8. Fluctuation of space-time and elementary particles[5]

The variety of distribution function $p(a)$ may be expected to be connected with the variety of elementary particles. To advance this idea a step forward, it might be hypothesized that the elementary particle is the very fluctuation of space-time itself.

Let us identify the probability distribution function $\pi(\xi)$, rewriting $p(a)$, with the absolute square of the internal state function $\varphi(\xi)$ of elementary particle,

$$\pi(\xi) = |\varphi(\xi)|^2, \tag{65}$$

$$\int_{-\infty}^{\infty} \pi(\xi)\, d\xi = 1, \tag{66}$$

and symmetralize τ' with respect to the initial and the final states,

$$\tau'^2 = \tau^2 - (\xi_2 - \xi_1)^2, \tag{67}$$

then, the modified Δ-functions will be given as follows:

$$\bar{\Delta}'(\tau, \kappa_m) = \iint_{-\infty}^{\infty} \pi(\xi_2)\, \bar{\Delta}(\tau', \kappa_m)\, \pi(\xi_1)\, d\xi_1\, d\xi_2 \tag{68}$$

and the similar relations for $\Delta^{(1)}$ and Δ-functions.

Another idea may be that $\varphi(\xi)$ itself, not $\pi(\xi)$, is used in the relation (68).

Generalizing the fluctuation to multi-dimensioal, if we take the three-dimensional harmonic oscillation as the internal state,

$$\varphi(\xi, \eta, \zeta) = (\kappa_0^{1/2}/\pi^{1/4})^3 \exp\left[-\kappa_0^2(\xi^2 + \eta^2 + \zeta^2)/2\right], \tag{69}$$

the propagation functions are shown to be properly regularized[6] and the modified static potential is given as below:

$$V_c' = (\sqrt{2}\,\kappa_0/\sqrt{\pi})(\kappa_0^2 r^2/4)\exp(\kappa_0^2 r^2/4)\left[K_1(\kappa_0^2 r^2/4) - K_0(\kappa_0^2 r^2/4)\right], \tag{70}$$

which is not singular and equal to $\sqrt{2}\,\kappa_0/\sqrt{\pi}$ at the origin and behaves like $1/r$ at infinity.

Closing remarks

(1) After the theory of non-local fields, Yukawa proposed the field theory of elementary domains,[17] of which space-time is supposed to be made of small domains, like cells. In his theory, the second quantization is formulated so as to satisfy the causality modified in the above.

(2) From the view-point of the distribution of space-time, Finsler spaces may be regarded as the superposition of Riemann spaces, corresponding to various values of directional variables.

(3) The theory of fields in Finsler spaces might be connected with the theory of string or membrane through the indicatrix.[18]

References

1) H. Yukawa: Phys. Rev. 77, 219 (1950); H. Yukawa: "Space-Time Description of Elementary Particles" in the Proceedings of the International Conference on Elementary Particles, Kyoto, 1965, p.139, where the references to preceding papers are found.

2) Y. Takano: Prog. Theor. Phys. 32, 365 (1964); 40, 1159 (1968); Lett. Nuovo Cimento 10, 747 (1974); 11, 486 (1974); "On the Theory of Fields in Finsler Spaces" in the Proceedings of the International Symposium on Relativity and Unified Field Theory, Calcutta, 1975-76, p.17; Lett. Nuovo Cimento, 35, 213 (1982); "Introduction to the Theory of Fields in Finsler Spaces" in Gravitational Measurements, Fundamental Metrology and Constants, edited by V. de Sabbata and V. N. Melnikov (Kluwer Academic, 1988), p.459; "General Survey of the Theory of Fields in Finsler Spaces" in the Memorial Volume for Akitsugu Kawaguchi, University of Athens (Athens, 1990).

3) H. Ishikawa: J. Math. Phys. 22, 995 (1981); Rep. Math. Phys. 25, 243 (1988).

4) S. Ikeda: J. Math. Phys. 26, 958 (1985).

5) G. S. Asanov: Finsler Geometry, Relativity and Gauge Theories (D. Reidel, 1985).

6) As to Finsler geometry, refer to the following books. P. Finsler: Über Kurven und Flächen in Allgemeiner Räumen, Dissertation, Göttingen (1918); E. Cartan: Les espace de Finsler, Actualités 79 (Hermann, 1934); Abh. aus dem Seminar für Vektor- und Tensoranalysis 4, 70 (1937); H. Rund: The Differential Geometry of Finsler Spaces (Springer, 1959); M. Matsumoto: Foundations of Finsler Geometry and Special Finsler Spaces (Kaiseisha, Otsu, Japan, 1986),

7) Y. Takano: Tensor, N.S., 40, 249 (1983); T. Ono and Y. Takano: Tensor, N.S., 49, 65, 253 (1990); Y. Takano and T. Ono: "The Differential Geometry of Spaces Whose Metric Tensor Depends on Spinor Variables and the Theory of Spinor Gauge Fields" in the Memorial Volume for Akitsugu Kawaguchi, University of Athens (Athens, 1990); Y. Takano and T. Ono: Tensor, N.S., 49, 269 (1990); T. Ono and Y. Takano: Tensor, N.S., 52, 56 (1993).

8) D. R. Yennie: Phys. Rev. 80, 1053 (1950); J. Rayski: Proc. Roy. Soc. A206, 575 (1951); O. Hara and H. Shimazu: Prog. Theor. Phys. 9, 137 (1953).

9) M. A. Markov: Nuclear Physics, 10, 140 (1959); A. A. Komar and M. A. Markov: Nuclear Physics, 12, 190 (1959).

10) Y. Takano: Prog. Theor. Phys. 26, 304 (1961); 27, 212 (1962).

11) H. Yukawa: "Extensions and Modifications of Quantum Field Theory" in the Proceedings of the Twelfth Conference on Physics, Brussels, 1961, p.235.

12) G. A. Armoudian, D. Ito and K. Mori: Nuovo Cimento, 38, 1895, 1903 (1965); Prog. Theor. Phys. Suppl. Extra Number, 626 (1965).

13) F. Bopp: Ann. d. Phys. 38, 345 (1940); 42, 537 (1942/43); S.

Sakata: Prog. Theor. Phys. 2, 30 (1947); A. Pais and G. E. Uhlenbeck: Phys. Rev. 79, 145 (1950).

14) R. P. Feynman: Phys. Rev. 74, 939, 1430 (1948).

15) Y. Takano: Prog. Theor. Phys. 38, 1185, 1186 (1967).

16) W. Pauli and F. Villars: Rev. Mod. Phys. 21, 434 (1949)

17) H. Yukawa: Prog. Theor. Phys. Suppl. Nos.37 & 38, 512 (1966); Y. Katayama and H. Yukawa: Prog. Theor. Phys. Suppl. No.41, 1 (1968); Y. Katayama, I. Umemura and H. Yukawa: Prog. Theor. Phys. Suppl. No.41, 22 (1968).

18) J. Rayski: Prog. Theor. Phys. 76, 1393 (1986).

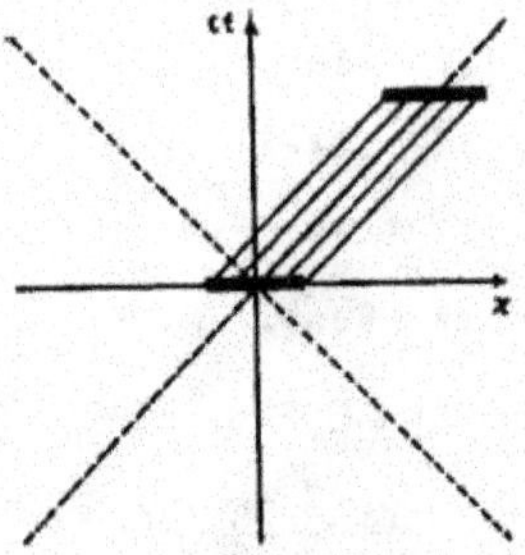

Fig. 1

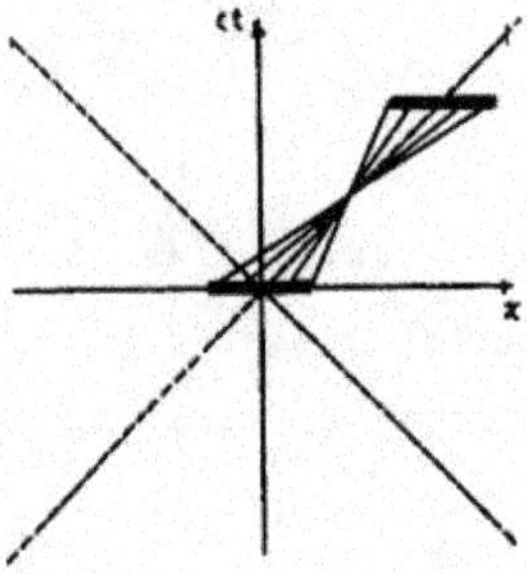

Fig. 2

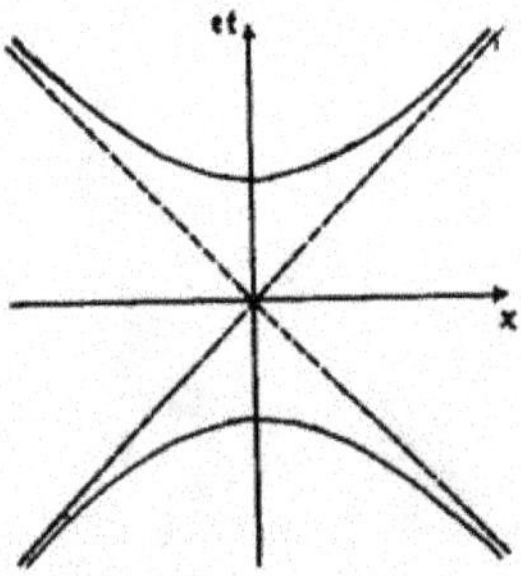

Fig. 3

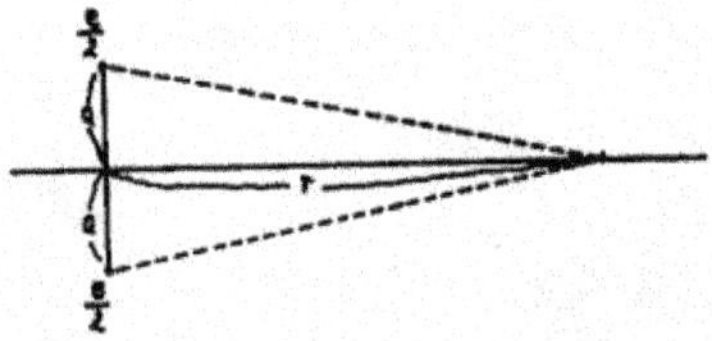

Fig. 4

Quantum Gravity Phenomenology

D.V. Nanopoulos
Center for Theoretical Physics, Dept. of Physics, Texas A&M University, College Station, TX 77843, USA

Astro Particle Physics Group, Houston Advanced Research Center (HARC), The Mitchell Campus, Woodlands, TX 77381, USA

Academy of Athens, Chair of Theoretical Physics, Division of Natural Sciences, 28 Panepistimiou Avenue, Athens 10679, Greece
E-mail: dimitri@soda.physics.tamu.edu

The possibility that quantum fluctuations in the structure of space-time at the Planck scale might be subject to experimental probes is discussed. The effects of space-time foam in an approach inspired by string theory, in which solitonic D-brane excitations are taken into account when considering the ground state are studied. The properties of this medium are described by analyzing the recoil of a D-particle which is induced by the scattering of a closed-string state. This recoil causes an energy-dependent perturbation of the background metric, which in turn induces an energy-dependent refractive index *in vacuo*, and stochastic fluctuations of the light cone. Distant astrophysical sources such as Gamma-Ray Bursters (GRBs) may be used to test this possibility, and an illustrative analysis of GRBs whose redshifts have been measured is presented. The propagation of massive particles through such a quantum space-time foam is also discussed.

1 Introduction

The concept of space time foam is an old one, first suggested by J.A. Wheeler[1], which has subsequently reappeared in various forms: see, for example[2,3,4,5,6,7,8]. The basic intuition is that quantum-gravitational fluctuations in the fabric of space-time cause the vacuum to behave like a stochastic medium. Most physicists who have studied the problem would surely agree that quantum-gravitational interactions must alter dramatically our classical perception of the space-time continuum when one attains Planckian energy scales $E \sim M_P \simeq 10^{19}$ GeV or Planckian distance scales $\ell \sim \ell_P \simeq 10^{-33}$ cm, which are the scales where gravitational interactions are expected to become strong. At issue are the following questions: how may classical space-time be altered at these scales? and is there any way of testing these possibilities?

At first sight, it might seem impossible to test such a suggestion within the foreseeable future, given the restrictions on the energies attainable with particle accelerators, and their corresponding limitations as microscopes. However, there are many instances in which new physics has revealed itself as a novel phenomenon far below its intrinsic mass scale M, a prime example being the weak interaction. In general, new-physics are suppressed by some inverse power of the heavy mass scale M, e.g., weak-interaction amplitudes are suppressed by $1/M_W^2$. However, some new-physics effects may be suppressed by just one power of the heavy mass scale, e.g., proton-decay amplitudes in some supersymmetric GUTs are $\propto 1/M_{GUT}$. Although gravitational amplitudes are generally suppressed by $1/M_P^2$, one should be open to the possibility that some quantum-gravitational effects might be suppressed simply by $1/M_P$. Moreover, there are many suggestions nowadays that M_P might not be a fundamental mass scale, and the quantum-gravitational effects might appear at some much lower scale related to the size(s) of one or more extra dimensions [9].

It has sometimes been suggested that Lorentz invariance might require quantum-gravitational effects to be suppressed by at least $\mathcal{O}[(E/M_P)^2]$, where E is a typical low-energy scale. However, Lorentz invariance is a casualty of many approaches to quantum gravity, and it is not clear how the very concept of space-time foam could be formulated Lorentz-invariantly. For example, many approaches to the physics of very small distances suggest that the classical space-time continuum may no longer exist, but might instead be replaced by a cellular structure.

String theory, plausibly in its current formulation as M theory, is at present the best (only?) candidate for a true quantum theory of gravity, so it is natural to ask what guidance it may offer us into the possible observability of quantum-gravitational effects. A first tool for this task was provided by two-dimensional string models [10], but a much more powerful tool has now been provided by D(irichlet) branes [11]. In this talk I review one particular D-brane approach to the modelling of space-time foam [12,13]. A characteristic feature of this approach is a treatment of D-brane recoil effects, in an attempt to incorporate the back-reaction of propagating particles on the ambient metric. This leads to the sacrifice of Lorentz invariance at the Planck scale, and suggests that space-time-foam effects arise already at the order $\mathcal{O}(E/M_P)$, in which case they might well be observable.

We have argued in the past that such minimally-suppressed quantum-gravitational effects could be probed in the neutral-kaon system [3], which is well known to be a sensitive laboratory for testing quantum mechanics and fundamental symmetries. In this talk, I focus more on the possibility that the propagation of other particles such as photons or neutrinos might be affected in a way that could be testable in the foreseeable future, for instance through astrophysical observations of pulsed sources such as γ-ray bursters (GRBs), active galactic nuclei (AGNs) or pulsars [5]. Our basic suggestion is that space-time foam may act as a non-trivial optical medium with, e.g., an energy- or frequency-dependent refractive index, through which the propagation of energetic particles might be slowed down so that they travel at less than the speed of light c: $\delta c/c \sim -E/M_P$. A secondary effect might be a stochastic spread in the velocities of particles with identical energies.

The primary tool for measuring small variations δc in the velocity of light c is the variation in arrival time [5,14] $\delta t \simeq -(L/c)\ (\delta c/c)$ observed for a photon travelling a distance L. This clearly places a premium on observing sources whose emissions exhibit structure on short time scales $\Delta t \lesssim \delta t$ located at large distances. Some typical numbers for some astrophysical sources are shown in Table 1. The relevant photon property that could be correlated with velocity variations δc is its frequency ν, or equivalently its energy E, for which characteristic values are also listed in Table 1. As I discuss in more detail later, any such effect could be expected to increase with E, and the simplest possibility, for which there is some theoretical support, is that $\delta c \propto E/M$, where M is some high energy scale. In this case, the relevant figure of merit for observational tests is the combination

$$\frac{L \cdot E}{c \cdot \delta t} \tag{1}$$

which measures directly the experimental sensitivity to such a high energy scale M. This sensitivity is also listed in Table 1, where we see that our favoured astrophysical sources are potentially sensitive to M approaching $M_P \simeq 10^{19}$ GeV, the mass scale at which gravity becomes strong. Therefore, these astrophysical sources may begin to challenge any theory of quantum gravity that predicts such a linear dependence of δc on E. Alternatively, it could be that $\delta c/c \simeq (E/\tilde{M})^2$, in which case the appropriate figure of merit is $E\sqrt{L/c\delta t}$. In this case, astrophysical observations may be sensitive to $\tilde{M} \sim 10^{11}$ TeV.

Table 1: *Observational Sensitivities and Limits on M. The mass-scale parameter M is defined by $\delta c/c = E/M$. The question marks in the Table indicate uncertain observational inputs. Hard limits are indicated by inequality signs.*

Source	Distance	E	Δt	Sensitivity to M
GRB 920229 [5,15]	3000 Mpc (?)	200 keV	10^{-2} s	0.6×10^{16} GeV (?)
GRB 980425	40 Mpc	1.8 MeV	10^{-3} s (?)	0.7×10^{16} GeV (?)
GRB 920925c	40 Mpc (?)	200 TeV (?)	200 s	0.4×10^{19} GeV (?)
Mrk 421 [16]	100 Mpc	2 TeV	280 s	$> 7 \times 10^{16}$ GeV
Crab pulsar [17]	2.2 kpc	2 GeV	0.35 ms	$> 1.3 \times 10^{15}$ GeV
GRB 990123	5000 Mpc	4 MeV	1 s (?)	2×10^{15} GeV (?)

2 Space-time foam from D-particle recoil

The string-inspired prototype model of space-time foam is based on one particular treatment of D branes[12,13]. This model involves naturally the breaking of Lorentz covariance, as a sort of spontaneous breaking. The basic idea may be summarized as follows. In the modern view of string theory, D particles must be included in the consistent formulation of the ground-state vacuum configuration[11]. Consider a closed-string state propagating in a $(D+1)$-dimensional space-time, which impacts a very massive D(irichlet) particle embedded in this space-time. We argue that the scattering of the closed-string state on the D particle induces recoil of the latter, which distorts the surrounding space-time in a stochastic manner. From the point of view of the closed-string particle and any low-energy spectators, this is a non-equilibrium process, in which information is 'lost' during the recoil, being carried by recoiling D-brane degrees of freedom that are inaccessible to a low-energy observer. Thus, although the entire process is consistent with quantum-mechanical unitarity, the low-energy effective theory is characterized by information loss and entropy production. From a string-theory point of view, the loss of information is encoded in a deviation from conformal invariance of the relevant world-sheet σ model that describes the recoil, that is compensated by the introduction of a Liouville field [18], which in turn is identified with the target time in the approach [10] adopted here.

In the case of a D-brane string soliton, its recoil after interaction with a closed-string state[19] is characterized by a pair of logarithmic operators[20]:

$$C_\epsilon \sim \epsilon\Theta_\epsilon(t), \qquad D_\epsilon \sim t\Theta_\epsilon(t) \tag{2}$$

defined on the boundary $\partial\Sigma$ of the string world sheet. The operators (2) act as deformations of the conformal field theory on the world sheet: $u_i \int_{\partial\Sigma} \partial_n X^i D_\epsilon$ describes the shift of the D brane induced by the scattering, where u_i is its recoil velocity, and $y_i \int_{\partial\Sigma} \partial_n X^i C_\epsilon$ describes quantum fluctuations in the initial position y_i of the D particle. It has been shown[21] that energy-momentum is conserved during the recoil process:

$$u_i = k_1 - k_2, \tag{3}$$

where $k_1(k_2)$ is the momentum of the propagating closed-string state before (after) the recoil, as a result of the summation over world-sheet genera. Thus the result (3) is an exact result, as far as world-sheet perturbation theory is concerned. I also note that $u_i = g_s P_i$, where P_i is the momentum and g_s is the string coupling, which is assumed here to be weak enough to ensure that

D branes are very massive, with mass $M_D = 1/(\ell_s g_s)$, where ℓ_s is the string length.

The correct specification of the logarithmic pair (2) entails a regulating parameter $\epsilon \to 0^+$, which appears inside the $\Theta_\epsilon(t)$ operator: $\Theta_\epsilon(t) = \int \frac{d\omega}{2\pi}\frac{1}{\omega - i\epsilon}e^{i\omega t}$. In order to realize the logarithmic algebra between the operators C and D, one takes [19]:

$$\epsilon^{-2} \sim \mathrm{Log}\Lambda/a \equiv \alpha \tag{4}$$

where Λ (a) are infrared (ultraviolet) world-sheet cut-offs. The pertinent two-point functions then have the following form [19]:

$$\begin{aligned} &< C_\epsilon(z)C_\epsilon(0) > \sim 0 + O[\epsilon^2] \\ &< C_\epsilon(z)D_\epsilon(0) > \sim 1 \\ &< D_\epsilon(z)D_\epsilon(0) > \sim \frac{1}{\epsilon^2} - 2\eta \log|z/L|^2 \end{aligned} \tag{5}$$

up to an overall normalization factor, which is the logarithmic algebra [20] in the limit $\epsilon \to 0^+$, modulo the leading divergence in the $< D_\epsilon D_\epsilon >$ recoil correlator. This leading divergent term will be important for our subsequent analysis.

The recoil deformations of the $D0$ brane (2) are relevant deformations, in the sense of conformal field theory, with anomalous dimension $-\epsilon^2/2$. However [21], the velocity operator D_ϵ (2) becomes exactly marginal, in a world-sheet renormalization-group sense, when it is divided by ϵ, in which case the recoil velocity is renormalized [21]:

$$u_i \to \overline{u}_i \equiv u_i/\epsilon \tag{6}$$

and becomes exactly marginal, playing the rôle of the physical velocity of the recoiling D particle.

Although such a renormalization is compatible with global world-sheet scaling, local world-sheet scale (conformal) symmetry is broken by the non-marginal character of the deformations (2), and restoration of conformal invariance requires Liouville dressing [18]. To determine the effect of such dressing on the space-time geometry, it is essential to write [12,13] the boundary recoil deformation as a bulk world-sheet deformation

$$\int_{\partial\Sigma} d\tau \overline{u}_i X^0 \Theta_\epsilon(X^0)\partial_n X^i = \int_\Sigma d^2\sigma \partial_\alpha \left([\overline{u}_i X^0]\Theta_\epsilon(X^0)\partial_\alpha X^i\right) \tag{7}$$

where the $\overline{u}_i$ denote the renormalized recoil couplings (6), in the sense discussed in [21]. As I have already mentioned, the couplings (7) are marginal on a flat world sheet, and become marginal on a curved world sheet if one dresses [18] the bulk integrand with a factor $e^{\alpha_i \phi}$, where ϕ is the Liouville field and α_i is the gravitational conformal dimension. This is related to the flat-world-sheet anomalous dimension $-\epsilon^2/2$ of the recoil operator, viewed as a bulk world-sheet deformation, as follows [18]:

$$\alpha_i = -\frac{Q_b}{2} + \sqrt{\frac{Q_b^2}{4} + \frac{\epsilon^2}{2}} \tag{8}$$

where Q_b is the central-charge deficit of the bulk world-sheet theory. In the recoil problem at hand, as discussed in [22], $Q_b^2 \sim \epsilon^4/g_s^2$ for weak deformations. This yields $\alpha_i \sim -\epsilon$ to leading order in perturbation theory in ϵ, to which I restrict ourselves here.

I next remark that, as the analysis of [12] indicates, the X^0-dependent operator $\Theta_\epsilon(X^0)$ scales as follows with ϵ for $X^0 > 0$: $\Theta_\epsilon(X^0) \sim e^{-\epsilon X^0}\Theta(X^0)$, where $\Theta(X^0)$ is a Heaviside step function without any field content, evaluated in the limit $\epsilon \to 0^+$. The bulk deformations, therefore, yield the following σ-model terms:

$$\frac{1}{4\pi\ell_s^2}\epsilon \sum_{I=m+1}^{D-1} g_{Ii} X^I e^{\epsilon(\phi_{(0)} - X^I_{(0)})}\Theta(X^I_{(0)}) \int_\Sigma \partial^\alpha X^I \partial^\alpha y_i \tag{9}$$

where the subscripts (0) denote world-sheet zero modes.

When we interpret the Liouville zero mode $\phi_{(0)}$ as target time, $\phi_{(0)} \equiv X^0 = t$, the deformations (9) yield space-time metric deformations in a σ-model sense, which were interpreted in [12] as expressing the distortion of the space-time surrounding the recoiling D-brane soliton. For clarity, we now drop the subscripts (0) for the rest of this paper. The resulting space-time distortion is then described by the metric elements:

$$G_{ij} = \delta_{ij}, G_{00} = -1, G_{0i} = \epsilon(\epsilon y_i + \epsilon \overline{u}_i t)\Theta_\epsilon(t), \ i = 1, \ldots D-1 \tag{10}$$

where the suffix 0 denotes temporal (Liouville) components.

The transformation laws for the couplings y_i and u_i, which are conjugate to D_ϵ and C_ϵ, are

$$u_i \to u_i \quad , \quad y_i \to y_i + u_i t \tag{11}$$

These are consistent with the interpretations of u_i as the velocity after the scattering process and y_i as the spatial collective coordinates of the brane, if and only if the parameter ϵ^{-2} is identified with the target Minkowski time t for $t \gg 0$ after the collision:

$$\epsilon^{-2} \simeq t \tag{12}$$

I have assumed in this analysis that the velocity u_i is small, as is appropriate in the weak-coupling régime studied here. The D-brane σ-model formalism is completely relativistic, and I anticipate that a complete treatment beyond the one-loop order discussed here will incorporate correctly all relativistic effects, including Lorentz factors wherever appropriate.

In view of (12), one observes that for $t \gg 0$ the metric (10) becomes to leading order:

$$G_{ij} = \delta_{ij}, G_{00} = -1, G_{0i} \sim \overline{u}_i, \qquad i = 1, \ldots D-1 \tag{13}$$

which is constant in space-time. However, the effective metric depends on the energy content of the low-energy particle that scattered on the D–particle, because of momentum conservation during the recoil process (3). This energy dependence is the primary deviation from Lorentz invariance induced by the D–particle recoil.

3 Refractive Index in Vacuo

I now proceed to discuss possible phenomenological consequences of the above phenomena, starting with the propagation of photons and relativistic particles. The above discussion of recoil suggests that the space-time background should be regarded as a non-trivial. Light propagating through media with non-trivial optical properties may exhibit a frequency-dependent refractive index, namely a variation in the light velocity with photon energy. Another possibility is a difference between the velocities of light with different polarizations, namely birefringence, and a third is a diffusive spread in the apparent velocity of light for light of fixed energy (frequency). Within the framework described in the previous Section, the first [5] and third [13] effects have been derived via a formal approach based on a Born-Infeld Lagrangian using D-brane technology. The possibility of birefringence has been raised [23] within a canonical approach to quantum gravity, but I do not pursue such a possibility here. A different approach to light propagation has been taken in [24], where quantum-gravitational fluctuations in the light-cone have been calculated. Here I use this formalism together with the microscopic model background obtained in the previous Section to derive a non-trivial refractive index and a diffusive spread in the arrival times of photons of given frequency.

As commented earlier, one may expect Lorentz invariance to be broken in a generic theory of quantum gravity, and specifically in the recoil context discussed in the previous Section. In the context of string theory, violations of Lorentz invariance entail the exploration of non-critical string backgrounds, since Lorentz invariance is related to the conformal symmetry that is a property of critical strings. As I discussed in the previous Section, a general approach to the formulation of non-critical string theory involves introducing a Liouville field [18] as a conformal factor on the string world sheet, which has non-trivial dynamics and compensates the non-conformal behaviour of the string background, and I showed in the specific case of D- branes that their recoil after interaction with a closed-string state produces a local distortion of the surrounding space-time (10).

Viewed as a perturbation about a flat target space-time, the metric (10) implies that that the only non-zero components of $h_{\mu\nu}$ are:

$$h_{0i} = \epsilon^2 \overline{u}_i t \Theta_\epsilon(t) \tag{14}$$

in the case of D-brane recoil. I now consider light propagation along the x direction in the presence of a metric fluctuation h_{0x} (14) in flat space, along a null geodesic given by $(dt)^2 = (dx)^2 + 2h_{0x}dtdx$. For large times $t \sim \text{Log}\Lambda/a \sim \epsilon^{-2}$ [22], $h_{ox} \sim \overline{u}$, and thus I obtain

$$\frac{cdt}{dx} = \overline{u} + \sqrt{1+\overline{u}^2} \sim 1 + \overline{u} + \mathcal{O}\left(\overline{u}^2\right) \tag{15}$$

where the recoil velocity $\overline{u}$ is in the direction of the incoming light ray. Taking into account energy-momentum conservation in the recoil process, which has been derived in this formalism as mentioned previously, one has a typical order of magnitude $\overline{u}/c = \mathcal{O}(E/M_D c^2)$, where $M_D = g_s^{-1} M_s$ is the D-brane mass scale, with $M_s \equiv \ell_s^{-1}$. Hence (15) implies a subluminal energy-dependent velocity of light:

$$c(E)/c = 1 - \mathcal{O}\left(E/M_D c^2\right) \tag{16}$$

which corresponds to a *classical* refractive index. This appears because the metric perturbation (14) is energy-dependent, through its dependence on $\overline{u}$.

The subluminal velocity (16) induces a delay in the arrival of a photon of energy E propagating over a distance L of order:

$$(\Delta t)_r = \frac{L}{c}\mathcal{O}\left(\frac{E}{M_D c^2}\right) \tag{17}$$

This effect can be understood physically from the fact that the curvature of space-time induced by the recoil is $\overline{u}-$ and hence energy-dependent. This affects the paths of photons in such a way that more energetic photons see more curvature, and thus are delayed with respect to low-energy ones.

As recalled above, the absence of superluminal light propagation was found previously via the formalism of the Born-Infeld lagrangian dynamics of D branes [21,13]. Furthermore, the result (17) is in agreement with the analysis of [14,5], which was based on a more abstract analysis of Liouville strings. It is encouraging that this result appears also in this more conventional general-relativity approach [24], in which the underlying physics is quite transparent.

Also a diffusive spread in the apparent velocity of light for light of fixed energy frequency is expected to lead to a diffusive spread in the arrival times of the photons. I find a contribution

$$|(\Delta t)_{obs}| \simeq \mathcal{O}\left(\bar{g}_s \frac{E}{M_D c^2}\right) \frac{L}{c} \tag{18}$$

to the RMS fluctuation in arrival times. As expected, the *quantum* effect (18) is suppressed by a power of the string coupling constant, when compared with the classical refractive index effect (17). The result (18) was derived in [13] using the techniques of Liouville string theory, via the Born-Infeld lagrangian for the propagation of photons in the D-brane foam. It should be noted that the recoil-induced effect (18) is larger than the effects discussed in [24], which are related to metric perturbations associated with the squeezed coherent states relevant to particle creation in conventional local field theories.

4 Maxwell's Equations Revisited

I now consider [25] the effects of the recoil-induced space time (13), viewed as a 'mean field solution' of the D–brane-inspired quantum-gravity model, on the propagation of electromagnetic waves. Maxwell's equations in the background metric (13) in empty space can be written as [26]:

$$\begin{aligned} \nabla \cdot B = 0, \qquad & \nabla \times H - \frac{1}{c}\frac{\partial}{\partial t} D = 0, \\ \nabla \cdot D = 0, \qquad & \nabla \times E + \frac{1}{c}\frac{\partial}{\partial t} B = 0, \end{aligned} \tag{19}$$

where

$$D = \frac{E}{\sqrt{h}} + H \times \mathcal{G}, \qquad B = \frac{H}{\sqrt{h}} + \mathcal{G} \times E \tag{20}$$

Thus, there is a direct analogy with Maxwell's equations in a medium with $1/\sqrt{h}$ playing the rôle of the electric and magnetic permeability. In our case [12], $h = 1$, so one has the same permeability as the classical vacuum. In the case of the constant metric perturbation (13), after some elementary vector algebra and, appropriate use of the modified Maxwell's equations, the equations (19) read:

$$\nabla \cdot E + \overline{u} \cdot \frac{1}{c}\frac{\partial}{\partial t} E = 0$$

$$\begin{aligned}
&\nabla \times B - (1-\overline{u}^2)\frac{1}{c}\frac{\partial}{\partial t}E + \overline{u} \times \frac{1}{c}\frac{\partial}{\partial t}B + (\overline{u}\cdot\nabla)E = 0\\
&\nabla \cdot B = 0\\
&\nabla \times E + \frac{1}{c}\frac{\partial}{\partial t}B = 0
\end{aligned} \tag{21}$$

Dropping non-leading terms of order $\overline{u}^2$ from these equations, one obtains after some straightforward algebra the following modified wave equations for E and B:

$$\begin{aligned}
&\frac{1}{c^2}\frac{\partial^2}{\partial^2 t}B - \nabla^2 B - 2(\overline{u}.\nabla)\frac{1}{c}\frac{\partial}{\partial t}B = 0\\
&\frac{1}{c^2}\frac{\partial^2}{\partial^2 t}E - \nabla^2 E - 2(\overline{u}.\nabla)\frac{1}{c}\frac{\partial}{\partial t}E = 0
\end{aligned} \tag{22}$$

If we consider one-dimensional motion along the x direction, we see that these equations admit wave solutions of the form

$$E_x = E_z = 0,\ E_y(x,t) = E_0 e^{i(kx-\omega t)},\ \ B_x = B_y = 0,\ B_z(x,t) = B_0 e^{i(kx-\omega t)}, \tag{23}$$

with the modified dispersion relation:

$$k^2 - \omega^2 - 2\overline{u}k\omega = 0 \tag{24}$$

Since the sign of $\overline{u}$ is that of the momentum vector k along the x direction, the dispersion relation (24) corresponds to *subluminal* propagation with a refractive index:

$$c(E) = c(1-\overline{u}) + \mathcal{O}(\overline{u}^2) \tag{25}$$

where we estimate that

$$\overline{u} = \mathcal{O}\left(\frac{E}{M_D c^2}\right) \tag{26}$$

with M_D the D-particle mass scale. This is in turn given by $M_D = g_s^{-1} M_s$ in a string model, where g_s is the string coupling and M_s is the string scale [13]. The relation (26) between $\overline{u}$ and the photon energy has been shown [21] to follow from a rigorous world-sheet analysis of modular divergences in string theory, but the details need not concern us here. It merely expresses elementary energy-momentum conservation, as discussed earlier.

The refractive index effect (25) is a mean-field effect, which implies a delay in the arrival times of photons, relative to that of an idealized low-energy photon for which quantum-gravity effects can be ignored, of order:

$$\Delta t \sim \frac{L}{c}|\overline{u}| = \mathcal{O}\left(\frac{EL}{M_D c^3}\right) \tag{27}$$

Figure 1: *Time distribution of the number of photons observed by BATSE in Channels 1 and 3 for GRB 970508, compared with the following fitting functions*[25]*: (a) Gaussian, (b) Lorentzian, (c) 'tail' function, and (d) 'pulse' function. We list below each panel the positions t_p and widths σ_p (with statistical errors) found for each peak in each fit. I recall that the BATSE data are binned in periods of 1.024 s.*

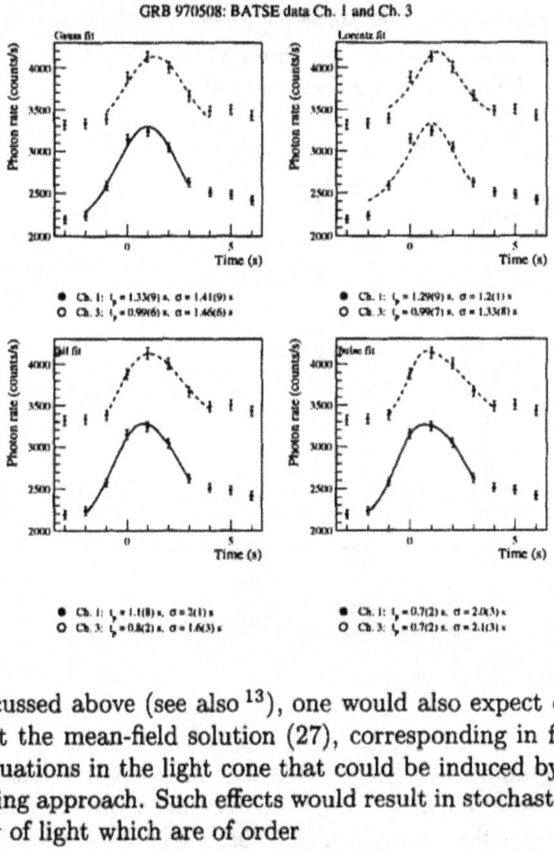

As I have discussed above (see also [13]), one would also expect quantum fluctuations about the mean-field solution (27), corresponding in field theory to quantum fluctuations in the light cone that could be induced by higher-genus effects in a string approach. Such effects would result in stochastic fluctuations in the velocity of light which are of order

$$\delta c \sim 8 g_s E / M_D c^2, \tag{28}$$

where g_s is the string coupling, which varies between $\mathcal{O}(1)$ and $\ll 1$ in different string models. Such an effect would motivate the following parametrization of any possible stochastic spread in photon arrival times:

$$(\delta \Delta t) = \frac{LE}{c\Lambda} \tag{29}$$

Figure 2: *Values of the shifts* $(\Delta t_p)^f$ *in the timings of the peaks fitted for each GRB studied using BATSE and OSSE data*^{2}5, *plotted versus* $\tilde{z} = 1 - (1+z)^{-1/2}$, *where* z *is the redshift. The indicated errors are the statistical errors in the 'pulse' fits provided by the fitting routine, combined with systematic error estimates obtained by comparing the results obtained using the 'tail' fitting function. The values obtained by comparing OSSE with BATSE Channel 3 data have been rescaled by the factor* $(E_{min}^{BATSE\ Ch.\ 3} - E_{max}^{BATSE\ Ch.\ 1})/(E_{min}^{OSSE} - E_{max}^{BATSE\ Ch.\ 3})$, *so as to make them directly comparable with the comparisons of BATSE Channels 1 and 3. The solid line is the best linear fit.*

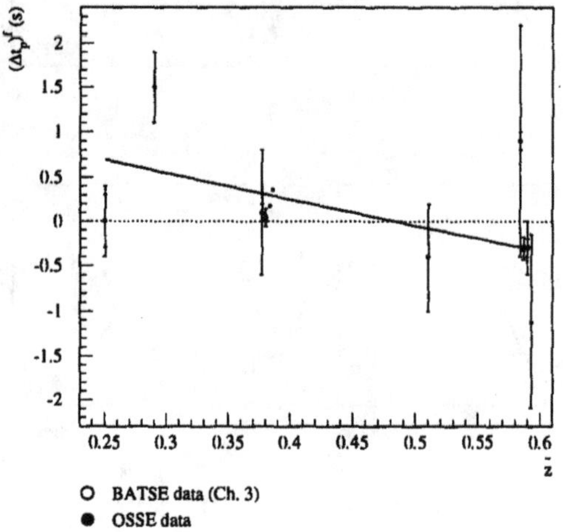

where the string approach suggests that $\Lambda \sim M_D c^2/8g_s$. I emphasize that, in contrast to the variation (25) in the refractive index - which refers to photons of different energy - the fluctuation (29) characterizes the statistical spread in the velocities of photons *of the same energy*. I recall that the stochastic effect (29) is suppressed, as compared to the refractive index mean field effect (26), by an extra power of g_s.

We presented in[25] a detailed analysis of the astrophysical data for a sample of Gamma Ray Bursters (GRB) whose redshifts z are known (see Fig. 1 for the data of a typical burst: GRB 970508). We looked (without success) for a correlation with the redshift, calculating a regression measure (see Fig. 2) for the effect (27) and its stochastic counterpart (29). Specifically, we looked for linear dependences of the 'observed' $\Delta t/\Delta E_0$ and the spread $\Delta\sigma/E$ on $\tilde{z} \equiv 2 \cdot [1 - (1/(1+z))^{1/2}] \simeq z - (3/4)z^2 + \ldots$. We determined limits on the quantum gravity scales M and Λ by constraining the possible magnitudes of the slopes in linear-regression analyses of the differences between the arrival times and widths of pulses in different energy ranges from five GRBs with measured redshifts, as functions of $\tilde{z}$. Using the current value for the Hubble expansion parameter, $H_0 = 100 \cdot h_0$ km/s/Mpc, where $0.6 < h_0 < 0.8$, we obtained the following limits [25]

$$M \gtrsim 10^{15}\ \text{GeV}, \quad \Lambda \gtrsim 2 \times 10^{15}\ \text{GeV} \tag{30}$$

on the possible quantum-gravity effects.

5 Massive Relativistic Particles Revisitied

So far, I have concentrated my attention on massless particles, namely photons, because of the very interesting experimental/observational possibilities they provide. It should however be clear, that the metric perturbation (13) produced by our D-brane recoil model, which implies a breakdown of Lorentz invariance, alters the Einstein dispersion relation for massive particles too. One expects, on general grounds, that the photon dispersion relation,

$$\omega^2 - k^2 + 2\bar{u}k\omega = 0 \tag{31}$$

will become in the case of massive particles,

$$\omega^2 - k^2 + 2\bar{u}k\omega - m_o^2 = 0\,, \tag{32}$$

leading to the modified Einstein relation,

$$\omega^2 \approx (k^2 + m_o^2)\left[1 - \frac{k^2}{\sqrt{k^2+m_o^2}}\frac{1}{M}\right]^2 \tag{33}$$

to leading order in $\mathcal{O}\left(\frac{k}{M}\right)$. It is useful to recast (33) in the more familiar notation:

$$E^2 = (p^2 + m_o^2)\left[1 - \left(\frac{1}{\sqrt{1+\frac{m_o^2}{p^2}}}\right)\frac{p}{M}\right]^2, \tag{34}$$

which in the non-relativistic limit yields:

$$E = m_o + \frac{p^2}{2m_o} - \frac{p^2}{M} + \cdots \mathcal{O}\left(\left(\frac{p}{M}\right)^2\right), \tag{35}$$

Thus, an effective rest mass

$$(m_{eff})_o \approx m_o \left(1 + \frac{2m_o}{M}\right). \tag{36}$$

appears in the non-relativistic kinetic energy: $E_{kin} = \frac{p^2}{2m_{eff}}$. Alternatively, one may use (34) to recast the T-shirt formula $E = mc^2$ in the form

$$E \approx mc^2 \left[1 - \frac{m}{M}\left(\frac{\overline{u}^2}{c^2}\right)\right], \tag{37}$$

where $m \equiv \frac{m_o}{\sqrt{1-\overline{u}^2/c^2}}$, as usual.

We see again that although quantum gravitational fluctuations in space-time, as modelled by D-brane quantum recoil, may lead to a spontaneous breakdown of Lorentz symmetry, the resulting corrections to the standard Einstein relations are very small, so that conventional Special Relativity is still a good approximation to the world.

I recall that another possible probe of quantum-gravitational effects on massive particles is offered by tests of quantum mechanics in the neutral kaon system. A parametrization of possible deviations from the Schrödinger equation has been given [3], assuming energy and probability conservation, in terms of quantities α, β, γ that must obey the conditions

$$\alpha, \gamma > 0, \qquad \alpha\gamma > \beta^2 \tag{38}$$

stemming from the positivity of the density matrix ρ. These parameters induce quantum decoherence and violate CPT [27]. Experimental data on neutral kaon decays so far agree perfectly with conventional quantum mechanics, imposing only the following upper limits [28]:

$$\alpha < 4.0 \times 10^{-17}\mathrm{GeV}, \qquad \beta < 2.3 \times 10^{-19}\mathrm{GeV}, \qquad \gamma < 3.7 \times 10^{-21}\mathrm{GeV} \tag{39}$$

I cannot help being impressed that these bounds are in the ballpark of m_K^2/M_P, which is the maximum magnitude that I could expect any such effect to have.

This and the example of photon propagation give hope that experiments may be able to probe physics close to the Planck scale, if its effects are suppressed by only one power of $M_P \simeq 10^{19}$ GeV. One should not exclude the possibility of being able to test some of the speculative ideas about quantum gravity reviewed in this article. Indeed, *if* the analysis of photon propagation can be extended to energetic neutrinos, and *if* GRBs emit ~ 10 s ν pulses, one could be sensitive to mass scales as large as 10^{28} GeV!

6 Conclusions

I have discussed here some possible low-energy probes of quantum gravity, concentrating on the possibility that the velocity of light might depend on its frequency, i.e., the corresponding photon energy. This idea is very speculative, and the model calculations that I have reviewed require justification and refinement. However, I feel that the suggestion is well motivated by the basic fact that gravity abhors rigid bodies, and the related intuition that the vacuum should exhibit back-reaction effects and act as a non-trivial medium. I recall that these features have appeared in several approaches to quantum gravity, including the canonical approach and ideas based on extra dimensions. Therefore, I consider the motivation from fundamental physics for a frequency-dependent velocity of light, and the potential significance of any possible observation, to be sufficient to examine this possibility from a phenomenological point of view.

As could be expected, we have found no significant effect in the data available on GRBs [25], either in the possible delay times of photons of higher energies, or in the possible stochastic spreads of velocities of photons with the same energy. However, it has been established that such probes may be sensitive to scales approaching the Planck mass, if these effects are linear in the photon energy. We expect that the redshifts of many more GRBs will become known in the near future, as alerts and follow-up observations become more effective, for example after the launch of the HETE II satellite [29,30]. Observations of higher-energy photons from GRBs would be very valuable, since they would provide a longer lever arm in the search for energy-dependent effects on photon propagation. Such higher-energy observations could be provided by future space experiments such as AMS [31] and GLAST [32].

This is not the only way in which quantum gravity might be probed: an example that we have advertized previously is provided by tests of quantum mechanics in the neutral-kaon system [3,27,28]. Forthcoming data from the DAΦNE accelerator may provide new opportunities for this quest. An alternative possibility might be provided by inteferrometric devices intended to detect gravity waves [8]. We also regard the emerging astrophysical suggestion of non-vanishing cosmological vacuum energy as a great opportunity for theoretical physics. If confirmed, this would provide a number to calculate in a complete quantum theory of gravity. The possibility that this vacuum energy might not be constant, but might actually be relaxing towards zero [33,34], is a possibility that may be tested by forthcoming cosmological observations. I therefore believe that the phrase 'experimental quantum gravity' may not be an oxymoron.

Acknowledgements

This work is supported in part by D.O.E. grant DE-F-G03-95-ER-40917.

1. J.A. Wheeler, Annals pf Physics (NY) **2**, 605 (1957); also in *Relativity, Groups and Topology*, eds. B.S. and C.M. de Witt (Gordon and Breach, New York, 1963).
2. S.W. Hawking, Nuclear Physics **B144**, 349 (1978); S.W. Hawking, D.N. Page, and C.N. Pope, Physics Letters **86B**, 175 (1979).
3. J. Ellis, J. Hagelin, D. Nanopoulos and M. Srednicki, Nuclear Physics **B241**, 381 (1984).
4. J. Ellis, N. E. Mavromatos and D. V. Nanopoulos, Physics Letters **B293**, 37 (1992).
5. G. Amelino–Camelia, J. Ellis, N. E. Mavromatos, D. V. Nanopoulos and S. Sarkar, Nature **393**, 763 (1998).
6. L. J. Garay, Physical Review **D58**, 124015 (1998).
7. A. Ashtekar, gr–qc/9901023.
8. G. Amelino–Camelia, Nature **398**, 216 (1999); gr-qc/9903080.
9. For a review see: I. Antoniadis, hep-th/9909212.
10. J. Ellis, N.E. Mavromatos and D.V. Nanopoulos, Physics Letters **B293** (1992), 37; *Lectures presented at the Erice Summer School, 31st Course: From Supersymmetry to the Origin of Space-Time*, Ettore Majorana Centre, Erice, 1993, (World Scientific, Singapore, 1994), p.1, hep-th/9403133.
11. J. Dai, R.G. Leigh and J. Polchinski, Modern Physics Letters **A4**, 2073 (1989); J. Polchinski, Physical Review Letters **75**, 184 (1995); hep-th/9611050.
12. J. Ellis, N. E. Mavromatos and D. V. Nanopoulos, International Journal of Modern Physics **A12**, 2639 (1997); *ibid.* **A13**, 1059 (1998).
13. J. Ellis, N. E. Mavromatos and D. V. Nanopoulos, gr–qc/9904068, gr–qc/9906029, General Relativity and Gravitation, in press.
14. G. Amelino-Camelia, J. Ellis, N.E. Mavromatos and D.V. Nanopoulos, International Journal of Modern Physics **A12**, 607 (1997).
15. B.E. Schaefer, Physical Review Letters **82**, 4964 (1999).
16. S.D. Biller *et. al.*, Physical Review Letters **83**, 2108 (1999).
17. P. Kaaret, Astronomy and Astrophysics **345**, L3 (1999).
18. I. Antoniadis, C. Bachas, J. Ellis and D.V. Nanopoulos, Physics Letters **B211**, 383 (1988); Nuclear Physics **B328**, 117 (1989).
19. I. I. Kogan, N. E. Mavromatos and J. F. Wheater, Physics Letters **B387**, 483 (1996).
20. V. Gurarie, Nuclear Physics **B410**, 535 (1993); J.S. Caux, I.I. Kogan and A.M. Tsvelik, Nuclear Physics **B466**, 444 (1996); M.A.I. Flohr, Interna-

tional Journal of Modern Physics **A11**, 4147 (1996), M.R. Gaberdiel and H.G. Kausch, Nuclear Physics **B489**, **293** (1996); I.I. Kogan and N.E. Mavromatos, Physics Letters **B375**, 111 (1996); M.R. Rahimi–Tabar, A. Aghamohammadi and M. Khorrami, Nuclear Physics **B497**, 555 (1997); I.I. Kogan, A. Lewis and O.A. Soloviev, International Journal of Modern Physics **A13**, 1345 (1998).

21. N. E. Mavromatos and R. Szabo, Physical Review **D59**, 104018 (1999).
22. J. Ellis, P. Kanti, N. E. Mavromatos, D. V. Nanopoulos
23. R. Gambini and Pullin J., Physical Review **D59**, 124021 (1999).
24. L. Ford, Physical Review **D 51**, 1692 (1995); H. Yu and L. Ford, gr–qc/9904082; gr–qc/9907037.
25. J. Ellis, K. Farakos, N.E. Mavromatos, V.A. Mitsou and D.V. Nanopoulos, astro-ph/9907340.
26. L. D. Landau and E. M. Lifshitz, *Classical Theory of Fields* (Pergamon Press, 1975), Vol. 2, page 257.
27. J. Ellis, N.E. Mavromatos and D.V. Nanopoulos, Physics Letters **B293**, 142 (1992); J. Ellis, J. Lopez, N.E. Mavromatos, and D.V. Nanopoulos, Physical Review **D53**, 3846 (1996).
28. R. Adler, *et al.* (CPLEAR Collaboration), J. Ellis, J. Lopez, N.E. Mavromatos, and D.V. Nanopoulos, Physics Letters **B364**, 239 (1995).
29. G. Ricker, in *Proc. Conference on Gamma-Ray Bursts in the Afterglow Era*, Rome, November 1998, Astronomy and Astrophysics Supplement Series, to appear.
30. K. Hurley, astro-ph/9812393.
31. S. Ahlen *et al.*, AMS Collaboration, *Nuclear Instruments and Methods* **A350**, 351 (1994).
32. E. D. Bloom *et al.*, GLAST Team, *Proc. International Heidelberg Workshop on TeV Gamma-Ray Astrophysics*, eds. H.J. Volk and F.A. Aharonian (Kluwer, 1996), 109.
33. J. Lopez and D.V. Nanopoulos, *Mod. Phys. Lett* **A9**, 2755 (1994); *ibid.* **A11**, 1 (1996).
34. J. Ellis, N.E. Mavromatos and D.V. Nanopoulos, gr-qc/9810086.

HAWKING RADIATION IN STRING THEORY AND THE STRING PHASE OF BLACK HOLES

M. RAMON MEDRANO [1],[2] and N. SANCHEZ [2]

Abstract

The quantum string emission by Black Holes is computed in the framework of the "string analogue model" (or thermodynamical approach), which is well suited to combine QFT and string theory in curved backgrounds (particulary here, as black holes and strings posses intrinsic thermal features and temperatures). The QFT-Hawking temperature T_H is upper bounded by the string temperature T_S in the black hole background. The black hole emission spectrum is an incomplete gamma function of $(T_H - T_S)$. For $T_H \ll T_S$, it yields the QFT-Hawking emission. For $T_H \to T_S$, it shows highly massive string states dominate the emission and undergo a typical string phase transition to a *microscopic* "minimal" black hole of mass $M_{\rm min}$ or radius $r_{\rm min}$ (inversely proportional to T_S) and string temperature T_S.

The semiclassical QFT black hole (of mass M and temperature T_H) and the string black hole (of mass $M_{\rm min}$ and temperature T_S) are mapped one into another by a "Dual" transform which links classical/QFT and quantum string regimes.

The string back reaction effect (selfconsistent black hole solution of the semiclassical Einstein equations with mass M_+ (radius r_+) and temperature T_+) is computed. Both, the QFT and string black hole regimes are well defined and bounded: $r_{\rm min} \leq r_+ \leq r_S$, $M_{\rm min} \leq M_+ \leq M$, $T_H \leq T_+ \leq T_S$.

The string "minimal" black hole has a life time $\tau_{\rm min} \simeq \frac{k_B c}{G\hbar} T_S^{-3}$.

[1]Departamento de Fisica Teórica, Facultad de Ciencias Físicas, Universidad Complutense, E-28040, Madrid, Spain.
[2]Observatoire de Paris, Demirm (Laboratoire Associé au CNRS UA
336, Observatoire de Paris et Ecole Normale Supérieure), 61 Avenue de l'Observatoire, 75014 Paris, France.

1 INTRODUCTION AND RESULTS

In the context of Quantum Field Theory in curved spacetime, Black Holes have an intrinsic Hawking temperature Ref. [1] given by

$$T_H = \frac{\hbar c}{4\pi k_B}\frac{(D-3)}{r_S} \qquad\qquad r_S \equiv L_{cl}$$

r_S being the Schwarzchild's radius (classical length L_{cl}).

In the context of Quantum String Theory in curved spacetime, quantum strings in black hole spacetimes have an intrinsic temperature given by

$$T_S = \frac{\hbar c}{4\pi k_B}\frac{(D-3)}{L_q}\, , \; L_q = \frac{bL_S(D-3)}{4\pi}\, , \; L_S \equiv \sqrt{\frac{\hbar\alpha'}{c}}\, ,$$

which is the same as the string temperature in flat spacetime (See Ref. [2] and Section 3 in this paper).

The QFT-Hawking temperature T_H is a measure of the Compton length of the Black Hole, and thus, of its "quantum size", or quantum property in the semiclassical-QFT regime. The Compton length of a quantum string is a direct measure of its size L_q. The string temperature T_S is a measure of the string mass, and thus inversely proportional to L_q.

The $\mathcal{R}$ or "Dual" transform over a length introduced in Ref. [3] is given by:

$$\begin{aligned} \tilde{L}_{cl} &= \mathcal{R}L_{cl} &= L_q \\ \tilde{L}_q &= \mathcal{R}L_q &= L_{cl} \end{aligned}$$

Under the $\mathcal{R}$-operation:

$$\tilde{T}_H = T_S$$

and

$$\tilde{T}_S = T_H$$

The QFT-Hawking temperature and the string temperature in the black hole background are $\mathcal{R}$-Dual of each other. This is valid in all spacetime dimensions D, and is a generic feature of QFT and String theory in curved backgrounds, as we have shown this relation for the respectives QFT-Hawking temperature and string temperature in de Sitter space Ref. [3]. In fact, the $\mathcal{R}$-transform maps QFT and string domains or regimes.

In this paper, we investigate the issue of Hawking radiation and the back reaction effect on the black hole in the context of String Theory. In principle, this question should be properly addressed in the context of String Field Theory. On the lack of a tractable framework for it, we work here in the framework of the string analogue model (or thermodynamical approach). This is a suitable approach for cosmology and black holes in order to combine QFT and string study and to go further in the understanding of quantum gravity effects. The thermodynamical approach is particularly appropriated and natural for black holes, as Hawking radiation and the string gas [4,5] posses intrinsic thermal features and temperatures.

In this approach, the string is a collection of fields Φ_n coupled to the curved background, and whose masses m_n are given by the degenerate string mass spectrum in the curved space considered. Each field Φ_n appears as many times the degeneracy of the mass level $\rho(m)$. (Althrough the fields Φ_n do not interact among themselves, they do with the black hole background).

In black hole spacetimes, the mass spectrum of strings is the same as in flat spacetime Ref. [2], therefore the higher masses string spectrum satisfies

$$\rho(m) = \left(\sqrt{\frac{\alpha' c}{\hbar}}\, m\right)^{-a} e^{b\sqrt{\frac{\alpha' c}{\hbar}}\, m}$$

(a and b being constants, depending on the model, and on the number of space dimensions).

We consider the canonical partition function ($\ln Z$) for the higher excited quantum string states of open strings (which may be or may be not supersymmetric) in the asymptotic (flat) black hole region. The gas of strings is at thermal equilibrium with the black hole at the Hawking temperature T_H, it follows that the canonical partition function, [Eq. (9)] is well defined for Hawking temperatures satisfying the condition

$$T_H < T_S$$

T_S represents a maximal or critical value temperature. This limit implies a minimum horizon radius

$$r_{\min} = \frac{b(D-3)}{4\pi} L_S$$

and a minimal mass for the black hole (BH).

$$M_{\min} = \frac{c^2(D-2)}{16\pi G} A_{D-2}\, r_{\min}^{D-3}$$

$$\left(M_{\min}(D=4) = \frac{b}{8\pi G}\sqrt{\hbar c^3 \alpha'}\right)$$

We compute the thermal quantum string emission of very massive particles by a D dimensional Schwarzschild BH. This highly massive emission, corresponding to the higher states of the string mass spectrum, is naturally expected in the last stages of BH evaporation.

In the context of QFT, BH emit particles with a Planckian (thermal) spectrum at temperature T_H. The quantum BH emission is related to the classical absorption cross section through the Hawking formula Ref. [1]:

$$\sigma_q(k, D) = \frac{\sigma_A(k, D)}{(e^{E(k)/k_B T_H} - 1)}$$

The classical total absorption spectrum $\sigma_A(k, D)$ Ref. [6] is entirely oscillatory as a function of the energy. This is exclusive to the black hole (other absorptive bodies do not show this property).

In the context of the string analogue model, the quantum emission by the BH is given by

$$\sigma_{\text{string}}(D) = \sqrt{\frac{\alpha' c}{\hbar}} \int_{m_0}^{\infty} \sigma_q(m, D)\, \rho(m) dm$$

$\sigma_q(m, D)$ being the quantum emission for an individual quantum field with mass m in the string mass spectrum. m_0 is the lowest mass from which the asymptotic expression for $\rho(m)$ is still valid.

We find $\sigma_{\text{string}}(D)$ as given by [Eq. (34)] (open strings). It consists of two terms: the first term is characteristic of a quantum thermal string regime, dominant for T_H close to T_S; the second term, in terms of the exponential-integral function E_i is dominant for $T_H \ll T_S$ from which the QFT Hawking radiation is recovered.
For $T_H \ll T_S$ (semiclassical QFT regime):

$$\sigma_{\text{string}}^{(\text{open})} \simeq B(D) \beta_H^{\frac{(D-5)}{2}}\, \beta_S^{-\frac{(D-3)}{2}}\, e^{-\beta_H m_0 c^2}\,, \ \beta_H \equiv (T_H k_B)^{-1}$$

For $T_H \to T_S$ (quantum string regime):

$$\sigma_{\rm string} \simeq B(D)\frac{1}{(\beta_H - \beta_S)}\ , \ \beta_S \equiv (T_S k_B)^{-1}$$

$B(D)$ is a precise computed coefficient [Eq. (34.a)].

The computed $\sigma_{\rm string}(D)$ shows the following: At the first stages, the BH emission is in the lighter particle masses at the Hawking temperature T_H as described by the semiclassical QFT regime (second term in [Eq. (34)]. As evaporation proceeds, the temperature increases, the BH radiates the higher massive particles in the string regime (as described by the first term of [Eq. (34)]). For $T_H \to T_S$, the BH enters its quantum string regime $r_S \to r_{\rm min}$, $M \to M_{min}$.

That is, "the BH becomes a string", in fact it is more than that, as [Eq. (34)] accounts for the back reaction effect too: The first term is characteristic of a Hagedorn's type singularity Ref. [5], and the partition function here has the same behaviour as this term. Its meaning is the following: At the late stages, the emitted BH radiation (highly massive string gas) dominates and undergoes a Carlitz's type phase transition Ref. [5] at the temperature T_S into a condensed finite energy state. Here such a state (almost all the energy concentrated in one object) is a *microscopic* (or "minimal") BH of size $r_{\rm min}$, (mass $M_{\rm min}$) and temperature T_S.

The last stage of the BH radiation, properly taken into account by string theory, makes such a phase transition possible. Here the T_S scale is in the Planck energy range and the transition is to a state of string size L_S. The precise detailed description of such phase transition and such final state deserve investigation.

A phase transition of this kind has been considered in Ref. [7]. Our results here supports and give a precise picture to some issues of BH evaporation discussed there in terms of purely thermodynamical considerations.

We also describe the (perturbative) back reaction effect in the framework of the semiclassical Einstein equations (c-number gravity coupled to quantum string matter) with the v.e.v. of the energy momentum tensor of the quantum string emission as a source. In the context of the analogue model, such stress tensor v.e.v. is given by:

$$\langle \tau_\mu^\nu(r)\rangle = \frac{\int_{m_0}^{\infty} \langle T_\mu^\nu(r,m)\rangle \sigma_q(m,D)\rho(m)dm}{\int_{m_0}^{\infty} \sigma_q(m,D)\rho(m)dm}$$

Where $< T_\mu^\nu(r,m) >$ is the v.e.v. of the QFT stress tensor of individual quantum fields of mass m in the higher excited string spectrum. The solution to the semiclassical Einstein equations is given by ([Eq. (53)], [Eq. (56)], [Eq. (60)]) ($D = 4$):

$$r_+ = r_S \left(1 - \frac{4}{21}\frac{\mathcal{A}}{r_S^6}\right)$$

$$M_+ = M \left(1 - \frac{4}{21}\frac{\mathcal{A}}{r_S^6}\right)$$

$$T_+ = T_H \left(1 + \frac{1}{3}\frac{\mathcal{A}}{r_S^6}\right)$$

The string form factor $\mathcal{A}$ is given by [Eq. (62)], it is finite and positive. For $T_H \ll T_S$, the back reaction effect in the QFT-Hawking regime is consistently recovered. Algebraic terms in $(T_H - T_S)$ are enterely stringly. In both cases, the relevant ratio $\mathcal{A}/r_S^6$ entering in the solution (r_+, M_+, T_+) is negligible. It is illustrative to show it in the two opposite regimes:

$$\left(\frac{\mathcal{A}}{r_S^6}\right)^{\text{open/closed}} \underset{T_H \ll T_S}{\simeq} \frac{1}{80640\pi}\left(\frac{M_{PL}}{M}\right)^4 \left(\frac{M_{PL}}{m_0}\right)^2 \ll 1$$

$$\left(\frac{\mathcal{A}}{r_{\text{min}}^6}\right)^{\text{closed}} \underset{T_H \to T_S}{\simeq} \frac{16}{7356}\left(\frac{\pi}{b}\right)^3 \left(\frac{M_S}{M_{PL}}\right)^2 \left(\frac{M_S}{m_0}\right)^2 \ll 1$$

M_{PL} being the Planck mass and $M_S = \frac{\hbar}{cL_S}$.

The string back reaction solution shows that the BH radius and mass decrease, and the BH temperature increases, as it should be. But here the BH radius is bounded from below (by r_{min} and the temperature does not blow up (as it is bounded by T_S). The "mass loss" and "time life" are:

$$-\left(\frac{dM}{dt}\right)_+ = -\left(\frac{dM}{dt}\right)\left(1 + \frac{20}{21}\frac{\mathcal{A}}{r_S^6}\right)$$

$$\tau_+ = \tau_H \left(1 - \frac{8}{7}\frac{\mathcal{A}}{r_S^6}\right)$$

The life time of the string black hole is $\tau_{\text{min}} = (\frac{K_{BC}}{G\hbar})T_S^{-3}$.

The string back reaction effect is finite and consistently describes both, the QFT regime (BH of mass M and temperature T_H) and the string regime (BH of mass M_{min} and temperature T_S). Both regimes are bounded as in string theory we have:

$$r_{\min} \le r_+ \le r_S \quad , \quad M_{\min} \le M_+ \le M$$

$$\tau_{\min} \le \tau_+ \le \tau_H \quad , \quad T_H \le T_+ \le T_S$$

The $\mathcal{R}$ "Dual" transform well summarizes the link between the two opposite well defined regimes: $T_H \ll T_S$ (ie $r_S \gg r_{\min}, M \gg M_{\min}$) and $T_H \to T_S$ (ie $r_S \to R_{\min}, M \to M_{\min}$).

This paper is organized as follows: In Section 2 we summarize the classical BH geometry and its semiclassical thermal properties in the QFT-Hawking regime. In Section 3 we derive the bonds imposed by string theory on this regime and show the Dual relation between the string and Hawking temperatures. In Section 4 we compute the quantum string emission by the BH. In Section 5 we compute its back reaction effect. Section 6 presents conclusions and remarks.

2 THE SCHWARZSCHILD BLACK HOLE SPACE TIME

The $D-$ dimensional Schwarzschild Black Hole metric reads

$$ds^2 = -a(r)c^2dt^2 + a^{-1}(r)dr^2 + r^2 d\,\Omega_{D-2}^2 \tag{1}$$

where

$$a(r) = 1 - \left(\frac{r_S}{r}\right)^{D-3} \tag{2}$$

being r_S the horizon (or Schwarzschild radius)

$$r_S = \left(\frac{16\pi GM}{c^2(D-2)A_{D-2}}\right)^{\frac{1}{D-3}} \tag{3.a}$$

and

$$A_{D-2} = \frac{2\pi^{\frac{(D-1)}{2}}}{\Gamma\left(\frac{(D-1)}{2}\right)} \tag{4}$$

(surface area per unit radius). G is the Newton gravitational constant. For $D = 4$ one has

$$r_S = \frac{2GM}{c^2} \tag{3.b}$$

The Schwarzschild Black Hole ($B.H$) is characterized by its mass M (angular momentum : $J = 0$; electric charge : $Q = 0$). The horizon [Eq. (3)] and the thermodynamical magnitudes associated to the $B.H$ – temperature (T), entropy (S), and specific heat (C_V) – are all expressed in terms of M (Table 1).

A brief review of these quantities is the following: in the context of QFT,Black Holes do emit thermal radiation at the Hawking temperature given by

$$T_H = \frac{\hbar\,\kappa}{2\pi\, k_B\, c} \tag{5.a}$$

where

$$\kappa = \frac{c^2(D-3)}{2r_S} \tag{5.b}$$

is the surface gravity. For $D = 4$,

$$T_H = \frac{\hbar\, c^3}{8\pi\, k_B\, GM} \tag{5.c}$$

The $B.H$ Entropy is proportional to the $B.H$ area $A = r_S^{D-2} A_{D-2}$ [Eq. (4)]

$$S = \frac{1}{4}\,\frac{k_B\, c^3}{G\,\hbar} A \tag{6}$$

and its specific heat $C_V = T\left(\frac{\partial S}{\partial T}\right)_V$ is negative

$$C_V = -\frac{(D-2)}{4}\,\frac{k_B\, c^3}{G\,\hbar}\, A_{D-2}\, r_S^{D-2} \tag{7.a}$$

In 4− dimensions it reads

$$C_V = -\frac{8\pi\, k_B\, G}{\hbar\, c}\, M^2 \tag{7.b}$$

As it is known, [Eq. (5)], [Eq. (6)] and [Eq. (7)] show that the $B.H-$ according to its specific heat being negative - increases its temperature in its quantum emission process (M decreases). Also, it could seem that, if the $B.H$ would evaporate completely ($M = 0$), the QFT-Hawking temperature T_H would become infinite. However, at this limit, and more precisely when $M \sim M_{PL}$, the fixed classical background approximation for the $B.H$ geometry breaks down, and the back reaction effect of the radiation matter on the $B.H$ must be taken into account. In Section 5, we will take into account this back reaction effect in the framework of string theory.

First, we will consider quantum strings in the fixed $B.H$ background. We will see that even in this approximation, quantum string theory not only can retard the catastrophic process but, furthermore, provides non-zero lower bounds for the $B.H$ mass (M) or horizon (r_S), and a finite (maximal) value for the $B.H$ temperature T_H as well.

	Dimension: D	$D=4$
r_S	$\left[\frac{16\pi GM}{c^2\,(D-2)\,A_{D-2}}\right]^{\frac{1}{D-3}}$	$\frac{2GM}{c^2}$
κ	$\frac{(D-3)\,c^2}{2r_S}$	$\frac{c^2}{2r_S}$
A	$A_{D-2}\, r_S^{D-2}$	$4\pi\, r_S^2$
T_H	$\frac{\hbar\,\kappa}{2\pi\, k_B\, c}$	$\frac{\hbar\, c}{4\pi\, k_B\, r_S}$
S	$\frac{1}{4}\,\frac{k_B\, c^3}{G\,\hbar}\, A$	$\frac{k_B\,\pi\, c^3}{G\,\hbar}\, r_S^2$
C_V	$-\frac{(D-2)}{4}\,\frac{k_B\, c^3}{G\,\hbar}\, A_{D-2}\, r_S^{D-2}$	$-\frac{2\pi\, k_B\, c^3}{G\,\hbar}\, r_S^2$

Table 1: Schwarzschild black hole thermodynamics. M ($B.H$ mass); r_S (Schwarzschild radius); κ (surface gravity); A (horizon area); T_H (Hawking temperature); S (entropy); C_V (specific heat); G and k_B (Newton and Boltzman constants); $A_{D-2} = 2\pi^{\frac{(D-1)}{2}}/\Gamma\left(\frac{(D-1)}{2}\right)$.

3 QUANTUM STRINGS IN THE BLACK HOLE SPACE TIME

The Schwarzschild black hole spacetime is asymptotically flat. Black hole evaporation - and any "slow down" of this process - will be measured by an observer which is at this asymptotic region. In Ref. [2] it has been found that the mass spectrum of quantum string states coincides with the one in Minkowski space. Critical dimensions are the same as well Ref. [2] ($D = 26$, open and closed bosonic strings; $D = 10$ super and heterotic strings).

Therefore, the asymptotic string mass density of levels in black hole spacetime will read as in Minkowski space

$$\rho(m) \sim \left(\sqrt{\frac{\alpha' c}{\hbar}}\, m\right)^{-a} e^{b\sqrt{\frac{\alpha' c}{\hbar}}\, m} \tag{8}$$

where $\alpha' \equiv \frac{c^2}{2\pi T}$ (T : string tension) has dimensions of (linear mass density)$^{-1}$; constants a/b depend on the dimensions and on the type of string Ref. [8]. For a non-compactified space-time these coefficients are given in Table 2.

Dimension	String Theory	a	b	$k_B T_S/c^2$
D	open bosonic	$(D-1)/2$	$2\pi\sqrt{\frac{D-2}{6}}$	$\left[2\pi\sqrt{\frac{(D-2)}{6}\left(\frac{\alpha' c}{\hbar}\right)}\right]^{-1}$
	closed	D		
26 (critical)	open bosonic	25/2	4π	$\left(4\pi\sqrt{\frac{\alpha' c}{\hbar}}\right)^{-1}$
	closed	26		
10 (critical)	open superstring	9/2	$\pi 2\sqrt{2}$	$\left(\pi 2\sqrt{2\left(\frac{\alpha' c}{\hbar}\right)}\right)^{-1}$
	closed superstring (type II)	10		
	Heterotic	10	$\pi(2+\sqrt{2})$	$\left[\pi(1+\sqrt{2})\sqrt{2\left(\frac{\alpha' c}{\hbar}\right)}\right]^{-1}$

Table 2: Density of mass levels $\rho(m) \sim m^{-a}\exp\{b\sqrt{\frac{\alpha' c}{\hbar}}m\}$. For open strings $\alpha'(\frac{c}{\hbar})m^2 \simeq n$; for closed strings $\alpha'(\frac{c}{\hbar})m^2 \simeq 4n$.

In this paper, strings in a $B.H$ spacetime are considered in the framework of the string analogue model. In this model, one considers the strings as a collection of quantum fields $\phi_1, \cdots, \phi_n$, whose masses are given by the string mass spectrum ($\alpha'(\frac{c}{\hbar})m^2 \simeq n$, for open strings and large n in flat spacetime). Each field of mass m appears as many times as the degeneracy of the mass level ; for higher excited modes this is described by $\rho(m)$ [Eq. (8)]. Although quantum fields do not interact among themselves, they do with the $B.H$ background.

In the asymptotic (flat) $B.H$ region, the thermodynamical behavior of the higher excited quantum string states of open strings, for example, is deduced from the canonical partition function Ref. [5]

$$\ln Z = \frac{V}{(2\pi)^d}\sqrt{\frac{\alpha' c}{\hbar}}\int_{m_0}^{\infty} dm\,\rho(m)\int d^d k \ln\left\{\frac{1+\exp\left[-\beta_H\left(m^2c^4+k^2\hbar^2c^2\right)^{\frac{1}{2}}\right]}{1-\exp\left[-\beta_H\left(m^2c^4+k^2\hbar^2c^2\right)^{\frac{1}{2}}\right]}\right\} \tag{9}$$

(d: number of spatial dimensions) where supersymmetry has been considered for the sake of generality; $\rho(m)$ is the asymptotic mass density given by [Eq. (8)]; $\beta_H = (k_B T_H)^{-1}$ where T_H is the $B.H$ Hawking temperature; m_0 is the lowest mass for which $\rho(m)$ is valid.

For the higher excited string modes, ie the masses of the $B.H$ and the higher string modes satisfy the condition

$$\beta_H\, m\, c^2 = \frac{4\pi\, m\, c}{(D-3)\,\hbar}\left[\frac{16\pi\, GM}{c^2(D-2)A_{D-2}}\right]^{\frac{1}{D-3}} \gg 1 \tag{10.a}$$

which reads for $D=4$

$$\beta_H\, m\, c^2 = \frac{8\pi\, GM\, m}{\hbar\, c} \gg 1 \tag{10.b}$$

(condition [Eq. (10.b)] will be considered later in section 4) the leading contribution to the r.h.s. of [Eq. (9)] will give as a canonical partition function

$$\ln Z \simeq \frac{2V_{D-1}\left(\alpha'\frac{c}{\hbar}\right)^{-\frac{(a-1)}{2}}}{\left(2\pi\,\beta_H\,\hbar^2\right)^{\frac{D-1}{2}}}\cdot\int_{m_0}^{\infty} d\,m\, m^{-a+\frac{D-1}{2}}\, e^{-(\beta_H-\beta_S)mc^2} \tag{11}$$

where $\beta_S = (k_B T_S)^{-1}$, being T_S [Eq. (8)]

$$T_S = \frac{c^2}{k_B\, b\left(\frac{\alpha' c}{\hbar}\right)^{\frac{1}{2}}} \tag{12}$$

the string temperature (Table 2). For open bosonic strings one divides by 2 the r.h.s. of [Eq. (11)] (leading contributions are the same for bosonic and fermionic sector).

From [Eq. (11)] we see that the definition of $\ln Z$ implies the following condition on the Hawking temperature

$$T_H < T_S \tag{13}$$

Furthermore, as T_H depends on the $B.H$ mass M, or on the horizon r_S , [Eq. (5.a)], [Eq. (5.b)] and [Eq. (3)], the above condition will lead to further conditions on the horizon. Then T_S represents a critical value temperature: $T_S \equiv T_{cr}$. In order to see this more clearly, we rewrite T_S in terms of the quantum string length scale

$$L_S = \left(\frac{\hbar\, \alpha'}{c}\right)^{\frac{1}{2}} \tag{14}$$

namely

$$T_S = \frac{\hbar\, c}{b\, k_B\, L_S} \tag{15}$$

From [Eq. (13)], and with the help of [Eq. (5.a)], [Eq. (5.b)] and [Eq. (15)], we deduce

$$r_S > \frac{b\,(D-3)}{4\pi}\, L_S \tag{16}$$

which shows that (first quantized) string theory provides a lower bound, or **minimum radius**, for the $B.H$ horizon.

Taking into account [Eq. (3)] and [Eq. (16)] we have the following condition on the $B.H$ mass

$$M > \frac{c^2\,(D-2) A_{D-2}}{16\pi\, G} \left[\frac{b\,(D-3)}{4\pi}\, L_S\right]^{D-3} \tag{17}$$

therefore there is a **minimal** $B.H$ **mass** given by string theory.

For $D = 4$ we have

$$r_S > \frac{b}{4\pi} L_S \tag{18}$$

$$M > \frac{c^2 b}{8\pi G} L_S \tag{19}$$

these lower bounds satisfy obviously [Eq. (3.b)]. [Eq. (19)] can be rewritten as

$$M > \frac{b M_{PL}^2}{8\pi M_S}$$

where $M_S = \frac{\hbar}{L_S c}$ is the string mass scale (L_S : reduced Compton wavelenght) and $M_{PL} \equiv \left(\frac{\hbar c}{G}\right)^{\frac{1}{2}}$ is the Planck mass. The minimal $B.H$ mass is then [Eq. (14)] and [Eq. (19)]

$$M_{\min} = \frac{b}{8\pi G} \sqrt{\hbar c^3 \alpha'}$$

It is appropriate, at this point, to make use of the $\mathcal{R}$ or Dual transformation over a length introduced in Ref. [3]. This operation is

$$\tilde{L}_{cl} = \mathcal{R} L_{cl} = \mathcal{L}_{\mathcal{R}} L_{cl}^{-1} = L_q \quad \text{and} \quad \tilde{L}_q = \mathcal{R} L_q = \mathcal{L}_{\mathcal{R}} L_q^{-1} = L_{cl} \tag{20.a}$$

where $\mathcal{L}_{\mathcal{R}}$ has dimensions of (lenght)^2; and it is given by $\mathcal{L}_{\mathcal{R}} = L_{cl} L_q$.

In our case, L_{cl} is the classical Schwarzschild radius, and $L_Q \equiv r_{\min} = (b(D-3)L_S)/4\pi$ [Eq. (16)].

The $\mathcal{R}$ transformation links classical lengths to quantum string lenghts, and more generally it links QFT and string theory domains Ref. [3].

and the string temperature is

$$T_H = \frac{\hbar c(D-3)}{4\pi k_B L_{cl}} \tag{20.b}$$

For the BH, the QFT-Hawking temperature is

$$T_S = \frac{\hbar c(D-3)}{4\pi k_B L_q} \tag{20.c}$$

Under the $\mathcal{R}$ operation we have

$$\tilde{T}_H = T_S \qquad \text{and} \qquad \tilde{T}_S = T_H \tag{20.d}$$

which are valid for all D.

From the above equations we can read as well

$$\tilde{T}_H \, \tilde{T}_S = T_S \, T_H$$

We see that under the $\mathcal{R}$-Dual operation, the QFT temperature and the string temperature in the BH background are mapped one into another. This appears to be a general feature for QFT and string theory in curved backgrounds, as we have already shown this relation in the de Sitter background Ref. [3].

It is interesting to express T_H and T_S in terms of their respective masses

$$\begin{aligned} T_H &= \tfrac{\hbar c(D-3)}{4\pi k_B} \left(\tfrac{16\pi G M}{c^2 (D-2) A_{D-2}}\right)^{-\frac{1}{D-3}} \\ T_H &= \tfrac{\hbar c^3}{8\pi k_B G M} \qquad\qquad (D=4) \end{aligned}$$

and

$$T_S = \frac{c^2 M_S}{b k_B}$$

4 THERMAL QUANTUM STRING EMISSION FOR A SCHWARZSCHILD BLACK HOLE

As it is known, thermal emission of massless particles by a black hole has been considered in the context of QFT Ref. [1], Ref. [9], Ref. [10]. Here, we are going to deal with thermal emission of high massive particles which correspond to the higher excited modes of a string. The study will be done in the framework of the string analogue model.

For a static $D-$ dimensional black hole, the quantum emission cross section $\sigma_q(k, D)$ is related to the total classical absorption cross section $\sigma_A(k, D)$ through the Hawking formula Ref. [1]

$$\sigma_q(k, D) = \frac{\sigma_A(k, D)}{e^{E(k)\beta_H} - 1} \tag{21}$$

where $E(k)$ is the energy of the particle (of momentum : $p = \hbar\, k$) and $\beta_H = (k_B T_H)^{-1}$, being T_H Hawking temperature [Eq. (5.a)] and [Eq. (5.b)].

The total absorption cross section $\sigma_A(k, D)$ in [Eq. (21)] has two terms Ref. [6], one is an isotropic $k-$ independent part, and the other has an oscillatory behavior, as a function of k, around the optical geometric constant value with decreasing amplitude and constant period. Here we will consider only the isotropic term, which is the more relevant in our case.

For a $D-$ dimensional black hole space-time, this is given by (see for example Ref.[2])

$$\sigma_A(k, D) = a(D)\, r_S^{D-2} \tag{22}$$

where r_S is the horizon [Eq. (3.a)] and [Eq. (4)] and

$$a(D) = \frac{\pi^{\frac{(D-2)}{2}}}{\Gamma\left(\frac{D-2}{2}\right)} \left(\frac{D-1}{D-3}\right) \left(\frac{D-1}{2}\right)^{\frac{2}{D-3}} \tag{23}$$

We notice that $\rho(m)$ [Eq. (8)] depends only on the mass, therefore we could consider, in our formalism, the emitted high mass spectrum as spinless. On the other hand, as we are dealing with a Schwarzschild black hole (angular momentum equal to zero), spin considerations can be overlooked. Emission is larger for spinless particles Ref. [11].

The number of scalar field particles of mass m emitted per unit time is

$$\langle n(m) \rangle = \int_0^\infty \langle\, n(k) \rangle\, d\mu(k) \tag{24}$$

where $d\mu(k)$ is the number of states between k and $k + dk$.

$$d\mu(k) = \frac{V_d}{(2\pi)^d} \frac{2\pi^{\frac{d}{2}}}{\Gamma\left(\frac{d}{2}\right)}\, k^{d-1}\, dk \tag{25}$$

and $\langle n(k)\rangle$ is now related to the quantum cross section σ_q ([Eq. (21)] and [Eq. (22)]) through the equation.

$$\langle n(k)\rangle = \frac{\sigma_q(k,D)}{r_S^{D-2}} \tag{26}$$

Considering the isotropic term for σ_q [Eq. (22)] and [Eq. (23)] we have

$$\langle n(k)\rangle = \frac{a(D)}{e^{E(k)\beta_H} - 1} \tag{27}$$

where $\beta_H = \frac{1}{k_B T_H}$, being T_H the BH temperature [Eq. (5)].

From [Eq. (24)] and [Eq. (27)], $\langle n(m)\rangle$ will be given by

$$\langle n(m)\rangle = F(D,\beta_H)\, m^{\frac{(D-3)}{2}}\, (mc^2\beta_H + 1)e^{-\beta_H mc^2} \tag{28}$$

where

$$F(D,\beta_H) \equiv \frac{V_{D-1}a(D)}{(2\pi)^{\frac{(D-1)}{2}}} \frac{(c^2)^{\frac{(D-3)}{2}}}{\beta_H^{\frac{(D+1)}{2}}\,(\hbar c)^{(D-1)}} \equiv A(D)\beta_H^{-\frac{D+1}{2}} \tag{29}$$

Large argument $\beta_H\, m\, c^2 \gg 1$, ie [Eq. (10.a)] and [Eq. (10.b)], and leading approximation have been considered in performing the k integral.

The quantum thermal emission cross section for particles of mass m is defined as

$$\sigma_q(m,D) = \int \sigma_q(k,D)d\mu(k) \tag{30.a}$$

and with the help of [Eq. (26)] we have

$$\sigma_q(m,D) = r_S^{D-2}\,\langle n(m)\rangle \tag{30.b}$$

where $\langle n(m)\rangle$ is given by [Eq. (28)].

In the string analogue model, the string quantum thermal emission by a BH will be given by the cross section

$$\sigma_{string}(D) = \sqrt{\frac{\alpha' c}{\hbar}} \int_{m_0}^{\infty} \sigma_q(m, D)\, \rho(m) dm \tag{31}$$

where $\rho(m)$ is given by [Eq. (8)], and $\sigma_q(m, D)$ by [Eq. (30)] and [Eq. (28)]; m_0 is the lowest string field mass for which the asymptotic value of the density of mass levels, $\rho(m)$, is valid.

For arbitrary D and a, we have from [Eq. (31)], [Eq. (30.b)], [Eq. (28)] and [Eq. (8)]

$$\sigma_{string}(D) = F(D, \beta_H) r_S^{D-2} \left(\sqrt{\frac{\alpha' c}{\hbar}}\right)^{-a+1} I_D(m, \beta_H - \beta_{cr}, a) \tag{32}$$

where $F(D, \beta_H)$ is given by [Eq. (29)] and

$$\beta_{cr} \equiv \beta_S = (k_B T_S)^{-1} = \frac{b}{c^2}\sqrt{\frac{\alpha' c}{\hbar}} \tag{33.a}$$

and

$$I_D(m, \beta_H - \beta_{cr}, a) \equiv \int_{m_0}^{\infty} m^{-a+\frac{D-3}{2}} \left(mc^2\beta_H + 1\right) e^{-(\beta_H - \beta_{cr})mc^2}\, dm \tag{33.b}$$

After a straightforward calculation we have

$$I_D(m, \beta_H - \beta_S, a) = \frac{c^2\beta_H}{[(\beta_H - \beta_S)c^2]^{-a+\frac{(D+1)}{2}}}\, \Gamma\left(-a + \frac{D+1}{2}, (\beta_H - \beta_S)c^2 m_0\right)$$
$$+ \frac{1}{[(\beta_H - \beta_S)c^2]^{-a+\frac{(D-1)}{2}}}\, \Gamma\left(-a + \frac{D-1}{2}, (\beta_H - \beta_S)c^2 m_0\right) \tag{33.c}$$

where $\Gamma(x, y)$ is the incomplete gamma function.

For open strings, $a = \frac{(D-1)}{2}$ [D : non-compact dimensions], we have

$$\sigma^{(\text{open})}_{\text{string}}(D) = A(D)\beta_H^{-\frac{(D+1)}{2}} r_S^{D-2} \left(\frac{c^2\beta_S}{b}\right)^{-\frac{(D-3)}{2}} \cdot \{\frac{\beta_H}{\beta_H-\beta_S} e^{-(\beta_H-\beta_S)c^2 m_0} - E_i\left(-(\beta_H-\beta_S)c^2 m_0\right)\} \quad (34)$$

where E_i is the exponential-integral function, and we have used [Eq. (29)] and [Eq. (33.A)].

When T_H approaches the limiting value T_S, and as $E_i(-x) \sim C + \ln x$ for small x, we have from [Eq. (34)]

$$\begin{aligned}
\sigma^{(\text{open})}_{\text{string}}(D) &= A(D)\beta_S^{-\frac{(D+1)}{2}} r_{\text{min}}^{D-2} \left(\tfrac{c^2}{b}\beta_S\right)^{-\frac{(D-3)}{2}} \\
&\quad \{\tfrac{\beta_S}{\beta_H-\beta_S} - C - \ln\left((\beta_H-\beta_S)m_0c^2\right)\} \\
&= B(D)\beta_S^{-1}\{\tfrac{\beta_S}{\beta_H-\beta_S} - C - \ln\left((\beta_H-\beta_S)c^2 m_0\right)\}
\end{aligned}$$

where

$$r_{\text{min}} = \frac{\hbar c(D-3)\beta_S}{4\pi}$$

and

$$B(D) \equiv A(D)\left(\frac{\hbar c(D-3)}{4\pi}\right)^{D-2}\left(\frac{c^2}{b}\right)^{-\frac{(D-3)}{2}} \quad (34.a)$$

For $\beta_H \to \beta_S$ the dominant term is

$$\sigma^{(\text{open})}_{\text{string}}(D) \overset{T_H \to T_S}{\simeq} B(D)\frac{1}{(\beta_H-\beta_S)} \quad (35.a)$$

for any dimension.

For $\beta_H \gg \beta_S$, ie $T_H \ll T_S$

$$\sigma^{(\text{open})}_{\text{string}}(D) \overset{T_H \ll T_S}{\simeq} A(D)\beta_H^{-\frac{(D+1)}{2}} r_S^{D-2}\left(\frac{c^2}{b}\beta_S\right)^{-\frac{(D-3)}{2}} e^{-\beta_H c^2 m_0}\left(1+\frac{1}{\beta_H c^2 m_0}\right)$$

$$\simeq B(D)\beta_H^{\frac{(D-5)}{2}}\,\beta_S^{-\frac{(D-3)}{2}}\,e^{-\beta_H c^2 m_0} \tag{35.b}$$

as $E_i(-x) \sim \frac{e^{-x}}{x} + \cdots$ for large x. For $D = 4$,

$$\sigma^{(\mathrm{open})}_{\mathrm{string}}(4) \simeq B(4)\left(\frac{1}{\beta_H\beta_S}\right)^{\frac{1}{2}} e^{-\beta_H c^2 m_0}$$

At this point, and in order to interpret the two different behaviours, we compare them with the corresponding behaviours for the partition function [Eq. (11)]. For open strings ($a = (D-1)/2$) $\ln Z$ is equal to

$$\ln Z_{\mathrm{open}} \simeq \frac{2V_{D-1}\left(\frac{\alpha' c}{\hbar}\right)^{-\frac{(D-3)}{4}}}{(2\pi\beta_H\hbar^2)^{\frac{(D-1)}{2}}} \cdot \frac{1}{(\beta_H-\beta_S)c^2} \cdot e^{-(\beta_H-\beta_S)m_0c^2}$$

For $\beta_H \to \beta_S$:

$$\ln Z_{\mathrm{open}} \underset{\beta_H \to \beta_S}{\simeq} \frac{2V_{D-1}\left(\frac{\alpha' c}{\hbar}\right)^{-\frac{(D-3)}{4}}}{(2\pi\beta_H\hbar^2)^{\frac{(D-1)}{2}}} \cdot \frac{1}{(\beta_H-\beta_S)c^2}$$

and for $\beta_H \gg \beta_S$:

$$\ln Z_{\mathrm{open}} \underset{\beta_H \gg \beta_S}{\simeq} \frac{2V_{D-1}\left(\frac{\alpha' c}{\hbar}\right)^{-\frac{(D-3)}{4}}}{(2\pi\hbar^2)^{\frac{(D-1)}{2}} c^2 \beta_H^{\frac{(D+1)}{2}}} \cdot e^{-\beta_H m_0 c^2}$$

The singular behaviour for $\beta_H \to \beta_S$, and all D, is typical of a string system with intrinsic Hagedorn temperature, and indicates a string phase transition (at $T = T_S$) to a condensed finite energy state (Ref. [5]). This would be the minimal black hole, of mass M_{min} and temperature T_S.

5 QUANTUM STRING BACK REACTION IN BLACK HOLE SPACE TIMES

When we consider quantised matter on a classical background, the dynamics can be described by the following Einstein equations

$$R_\mu^\nu - \frac{1}{2}\,\delta_\mu^\nu R = \frac{8\pi G}{c^4}\,\langle \tau_\mu^\nu \rangle \tag{35}$$

The space-time metric $g_{\mu\nu}$ generates a non-zero vacuum expectation value of the energy momentum tensor $\langle \tau_\mu^\nu \rangle$, which in turn, acting as a source, modifies the former background. This is the so-called back reaction problem, which is a semiclassical approach to the interaction between gravity and matter.

Our aim here is to study the back reaction effect of higher massive (open) string modes (described by $\rho(m)$, [Eq. (8)]) in black holes space-times. This will give us an insight on the last stage of black hole evaporation. Back reaction effects of massless quantum fields in these equations were already investigated Ref. [12], Ref. [13], Ref. [14].

As we are also interested in stablishing the differences, and partial analogies, between string theory and the usual quantum field theory for the back reaction effects in black holes space-times, we will consider a 4$-$ dimensional physical black hole.

The question now is how to write the appropriate energy-momentum tensor $\langle \tau_\mu^\nu \rangle$ for these higher excited string modes. For this purpose, we will consider the framework of the string analogue model. In the spirit of this model, the v.e.v. of the stress tensor $\langle \tau_\mu^\nu \rangle$ for the string higher excited modes is defined by

$$\langle \tau_\mu^\nu(r) \rangle = \frac{\int_{m_0}^{\infty} \langle T_\mu^\nu(r,m) \rangle \, \langle n(m) \rangle \rho(m)\, dm}{\int_{m_0}^{\infty} \langle n(m) \rangle \rho(m)\, dm} \tag{36}$$

where $\langle T_\mu^\nu(r,m) \rangle$ is the Hartle-Hawking vacuum expectation value of the stress tensor of an individual quantum field, and the r dependence of $\langle T_\mu^\nu \rangle$ preserves the central gravitational character of the problem ; $\rho(m)$ is the string mass density of levels [Eq. (8)] and $\langle n(m) \rangle$ is the number of field particles of mass m emitted per unit time, [Eq. (28)].

For a static spherically symmetric metric

$$ds^2 = g_{00}(r)c^2\,dt^2 + g_{rr}(r)dr^2 + r^2 d\Omega_2^2 \tag{37}$$

where $g_{00}(r) < 0\,(r > r_S)$ for a Schwarzschild black hole solution, the semiclassical Einstein equations [Eq. (36)] read

$$\frac{8\pi G}{c^4}\langle \tau_r^r\rangle = g_{rr}^{-1}\left(\frac{1}{r}\frac{d\ln g_{00}}{dr} + \frac{1}{r^2}\right) - \frac{1}{r^2} \tag{39.a}$$

$$\frac{8\pi G}{c^4}\langle \tau_0^0\rangle = g_{rr}^{-1}\left(\frac{1}{r^2} - \frac{1}{r}\frac{d\ln g_{rr}}{dr}\right) - \frac{1}{r^2} \tag{39.b}$$

For Schwarzschild boundary conditions

$$g_{00}(r_\infty)\, g_{rr}(r_\infty) = -1 \tag{40.a}$$

where

$$g_{rr}(r_\infty) = \left(1 - \frac{r_S}{r_\infty}\right)^{-1} \tag{40.b}$$

the solution to [Eq. (39.a)] and [Eq. (39.b)] is given by

$$g_{rr}^{-1}(r) = 1 - \frac{2GM}{c^2\,r} + \frac{8\pi G}{c^4\,r}\int_{r_\infty}^{r}\langle\tau_0^0(r')\rangle\, r'^2 dr' \tag{41.a}$$

$$g_{00}(r) = -g_{rr}^{-1}(r)\cdot\exp\left\{\frac{8\pi G}{c^4}\int_{r_\infty}^{r}\left(\langle\tau_r^r(r')\rangle - \langle\tau_0^0(r')\rangle\right) r' g_{rr}(r')dr'\right\} \tag{41.b}$$

where M is the black hole mass measured from r_∞ (r_∞ may be infinite or the radius of a cavity where the black hole is put inside to maintain the thermal equilibrium).

In order to write $\langle T_\mu^\nu(m,r)\rangle$ for an individual quantum field in the framework of the analogue model, we notice that $\rho(m)$ [Eq. (8)] depends only on m ; therefore, we will consider for simplicity the vacuum expectation value of the stress tensor for a massive scalar field.

For the Hartle-Hawking vacuum (black-body radiation at infinity in equilibrium with a black hole at the temperature T_H), and when the (reduced) Compton wave length of the massive particle

($\lambda = \frac{\hbar}{mc}$) is much smaller than the Schwarzschild radius (r_S)

$$\frac{\hbar\, c}{2GMm} \ll 1 \tag{42}$$

(same condition as the one of [Eq. (10.b)]), $\langle T^r_r \rangle$ and $\langle T^0_0 \rangle$ for the background $B.H$ metric ([Eq. (1)], $D = 4$) read Ref. [13]

$$\frac{8\pi G}{c^4} \langle T^r_r \rangle = \frac{A}{r^8} F_1\left(\frac{r_S}{r}\right) \tag{43.a}$$

$$\frac{8\pi G}{c^4} \langle T^0_0 \rangle = \frac{A}{r^8} F_2\left(\frac{r_S}{r}\right) \tag{43.b}$$

where

$$A = \frac{M^2\, L^6_{PL}}{1260\pi m^2} \tag{43.c}$$

$$F_1\left(\frac{r_S}{r}\right) = 441 - \zeta 2016 + \frac{r_S}{r}\,(-329 + \zeta 1512) + O(m^{-4}) \tag{43.d}$$

$$F_2\left(\frac{r_S}{r}\right) = -1125 + \zeta 5040 + \frac{r_S}{r}\,(1237 - \zeta 5544) + O(m^{-4}) \tag{43.e}$$

M and m are the black hole and the scalar field masses respectively, ζ (a numerical factor) is the scalar coupling parameter ($-\frac{\zeta R \phi^2}{2}$; R : scalar curvature, ϕ : scalar field) and $L_p \equiv (\frac{\hbar G}{c^3})^{\frac{1}{2}}$ is the Planck length.

From [Eq. (36)], [Eq. (43.a)] and [Eq. (43.b)] the v.e.v. of the string stress tensor will read

$$\frac{8\pi G}{c^4} \langle \tau^r_r \rangle = \frac{A}{r^8} F_1\left(\frac{r_S}{r}\right) \tag{44.a}$$

$$\frac{8\pi G}{c^4} \langle \tau^0_0 \rangle = \frac{A}{r^8} F_2\left(\frac{r_S}{r}\right) \tag{44.b}$$

where

$$\mathcal{A} = \frac{M^2 L_{PL}^6}{1260\pi} \frac{\int_{m_0}^{\infty} m^{-2} \langle n(m) \rangle \rho(m)\, dm}{\int_{m_0}^{\infty} \langle n(m) \rangle \rho(m)\, dm} \tag{45}$$

We return now to [Eq. (41.a)] and [Eq. (41.b)] which, with the help of [Eq. (44.a)], [Eq. (44.b)] and [Eq. (3.b)], can be rewritten as

$$g_{rr}^{-1}(r) = 1 - \frac{r_S}{r} + \frac{\mathcal{A}}{r} \int_{r_\infty}^{r} F_2\left(\frac{r_S}{r}\right) \frac{1}{r'^6}\, dr' \tag{46}$$

$$g_{00}(r) = -g_{rr}^{-1}(r) \cdot \exp\left\{\mathcal{A} \int_{r_\infty}^{r} \left[F_1\left(\frac{r_S}{r}\right) - F_2\left(\frac{r_S}{r}\right)\right] \frac{g_{rr}(r')}{r'^7}\, dr'\right\} \tag{47}$$

A Schwarzschild black body configuration

$$g_{00}(r) = -g_{rr}^{-1}(r) \tag{48}$$

is obtained when [Eq. (47)]

$$F_1\left(\frac{r_S}{r}\right) - F_2\left(\frac{r_S}{r}\right) \equiv (1566 - \zeta 7056)\left(1 - \frac{r_S}{r}\right) = 0 \tag{49}$$

ie for $\zeta = \frac{87}{392}$ for all r.

Then from [Eq. (46)] and [Eq. (43.e)], we obtain

$$g_{rr}^{-1} = 1 - \frac{r_S}{r} - \frac{\mathcal{A}}{21 r^6}\left[23\left(\frac{r_S}{r}\right) - 27\right] \tag{50}$$

From the above equation it is clear that the quantum matter back reaction modifies the horizon, r_+, which will be no longer equal to the classical Schwarzschild radius r_S. The new horizon will satisfy

$$g_{rr}^{-1} = 0 \tag{51.a}$$

ie

$$r_+^7 - r_S r_+^6 + \mathcal{A}\frac{27}{21} r_+ - \mathcal{A}\frac{23}{21} r_S = 0 \tag{51.b}$$

In the approximation we are dealing with ($O(m^{-4})$ ie $\mathcal{A}^2 \ll \mathcal{A}$), the solution will have the form

$$r_+ \simeq r_S(1+\epsilon)\,,\ \epsilon \ll 1 \tag{52}$$

From [Eq. (51.b)] we obtain

$$r_+ \simeq r_S\left(1 - \frac{4\mathcal{A}}{21 r_S^6}\right) \tag{53}$$

which shows that the horizon decreases.

Let us consider now the surface gravity, which is defined as

$$k(r_+) = \frac{c^2}{2}\frac{dg_{rr}^{-1}}{dr}|_{r=r_+} \tag{54}$$

(in the absence of back reaction, $k(r_+) = k(r_S)$ given by [Eq. (5.b)] for $D = 4$).

From [Eq. (50)], [Eq. (53)] and [Eq. (54)] we get

$$k(r_+) = \frac{c^2}{2r_S}\left(1 + \frac{1}{3}\frac{\mathcal{A}}{r_S^6}\right) \tag{55}$$

The black hole temperature will then be given by

$$T_+ = \frac{\hbar\kappa(r_+)}{2\pi k_B c} \simeq T_H\left(1 + \frac{1}{3}\frac{\mathcal{A}}{r_S^6}\right) \tag{56}$$

where $T_H = \frac{\hbar c}{4\pi k_B r_S}$

([Eq. (5.a)] and [Eq. (5.b)] for $D = 4$).The Black hole temperature increases due to the back reaction.

Due to the quantum emission the black hole suffers a loss of mass. The mass loss rate is given by a Stefan-Boltzman relation. Without back reaction, we have

$$-\left(\frac{dM}{dt}\right) = \sigma 4\pi r_S^2 T_H^4 \tag{57}$$

where σ is a constant.

When back reaction is considered, we will have

$$-\left(\frac{dM}{dt}\right)_+ = \sigma 4\pi r_+^2 T_+^4 \tag{58}$$

where r_+ is given by [Eq. (53)] and T_+ by [Eq. (56)]. Inserting these values into the above equation we obtain

$$-\left(\frac{dM}{dt}\right)_+ \simeq -\left(\frac{dM}{dt}\right)\left(1+\frac{20\mathcal{A}}{21r_S^6}\right) \tag{59}$$

On the other hand, the modified black hole mass is given by

$$M_+ \equiv \frac{c^2}{2G} r_+ \simeq M\left(1-\frac{4\mathcal{A}}{21r_S^6}\right) \tag{60}$$

which shows that the mass decreases.

From [Eq. (59)] and [Eq. (60)], we calculate the modified life time of the black hole due to the back reaction

$$\tau_+ \simeq \tau_H\left(1-\frac{8\mathcal{A}}{7r_S^6}\right) \tag{61}$$

We see that $\tau_+ < \tau_H$ since $\mathcal{A} > 0$.

We come back to the string back reaction "form factor" $\mathcal{A}$ [Eq. (45)] which can be rewritten as

$$\mathcal{A} = \frac{M^2 L_{PL}^6}{1260\pi} \cdot \frac{N}{De} \tag{62}$$

where

$$N = \int_{m_0}^{\infty} m^{-a+\frac{D-7}{2}} \left(mc^2\beta_H + 1\right) e^{-(\beta_H-\beta_S)mc^2}\, dm \tag{63}$$

and [Eq. (33)]

$$De = I_D(m, \beta_H - \beta_S, a) \tag{64}$$

where use of [Eq. (28)] and [Eq. (8)] has been made (common factors for numerator and denominator cancelled out).

For arbitrary D and a, N is given by

$$\begin{aligned} N &= \frac{c^2\beta_H}{[(\beta_H-\beta_S)c^2]^{-a+\frac{D-3}{2}}}\,\Gamma\left(-a+\frac{D-3}{2}, (\beta_H-\beta_S)c^2m_0\right) \\ &+ \frac{1}{[(\beta_H-\beta_S)c^2]^{-a+\frac{D-5}{2}}}\,\Gamma\left(-a+\frac{D-5}{2}, (\beta_H-\beta_S)c^2m_0\right) \end{aligned} \tag{65}$$

In particular, for open strings ($a = \frac{(D-1)}{2}$) we have for N and De

$$\begin{aligned} N &= c^4\beta_H(\beta_H-\beta_S)[E_i\left(-(\beta_H-\beta_S)m_0c^2\right) + \frac{e^{-(\beta_H-\beta_S)m_0c^2}}{(\beta_H-\beta_S)m_0c^2}] \\ &- \frac{(\beta_H-\beta_S)^2c^4}{2}\,[E_i\left(-(\beta_H-\beta_S)m_0c^2\right) + e^{-(\beta_H-\beta_S)m_0c^2} \\ &\cdot \left(\frac{1}{(\beta_H-\beta_S)m_0c^2} - \frac{1}{(\beta_H-\beta_S)^2m_0^2c^4}\right)] \end{aligned} \tag{66}$$

and

$$De = \frac{\beta_H}{\beta_H-\beta_S}\, e^{-(\beta_H-\beta_S)c^2m_0} - E_i\left(-(\beta_H-\beta_S)c^2m_0\right) \tag{67}$$

For $\beta_H \to \beta_S$ ($M \to M_{\min}, r_S \to r_{\min}$) we have for the open string form factor

$$\mathcal{A}_{\rm open} \simeq \frac{M_{\rm min}^2 L_{PL}^6 (\beta_H - \beta_S)}{1260\pi\, \beta_S} \left(\frac{1}{2m_0^2} + \frac{c^2 \beta_S}{m_0} \right) \tag{68}$$

Although the string analogue model is in the spirit of the canonical ensemble-all (higher) massive string fields are treated equally – we will consider too, for the sake of completeness, the string "form factor" $\mathcal{A}$ for closed strings.

For $a = D$ (D: non compact dimensions), from [Eq. (33.b)], [Eq. (33.c)] and [Eq. (64)], we have the following expressions

$$\begin{aligned} De &= I_D(m, \beta_H - \beta_S, D) \\ &= \frac{c^2 \beta_H}{[(\beta_H - \beta_S)c^2]^{-\frac{(D-1)}{2}}} \, \Gamma\left(-\frac{D-1}{2}, (\beta_H - \beta_S) m_0 c^2\right) \\ &+ \frac{1}{[(\beta_H - \beta_S)c^2]^{-\frac{(D+1)}{2}}} \, \Gamma\left(-\frac{D+1}{2}, (\beta_H - \beta_S) m_0 c^2\right) \end{aligned} \tag{69}$$

and [Eq. (63)]

$$\begin{aligned} N &= \frac{c^2 \beta_H}{[(\beta_H - \beta_S)c^2]^{-\frac{(D+3)}{2}}} \, \Gamma\left(-\frac{D+3}{2}, (\beta_H - \beta_S) c^2 m_0\right) \\ &+ \frac{1}{[(\beta_H - \beta_S)c^2]^{-\frac{(D+5)}{2}}} \, \Gamma\left(-\frac{D+5}{2}, (\beta_H - \beta_S) c^2 m_0\right) \end{aligned} \tag{70}$$

For $\beta_H \to \beta_S$ and D even, we have then

$$\begin{aligned} N &= c^2 \beta_S [(\beta_H - \beta_S)c^2]^{\frac{(D+3)}{2}} \, \Gamma\left(-\frac{D+3}{2}\right) + \frac{c^2 \beta_S}{(m_0)^{\frac{(D+3)}{2}} \left(\frac{D+3}{2}\right)} \\ &+ [(\beta_H - \beta_S)c^2]^{\frac{(D+5)}{2}} \, \Gamma\left(-\frac{D+5}{2}\right) + \frac{1}{\left(\frac{D+5}{2}\right) m_0^{\frac{(D+5)}{2}}} \end{aligned} \tag{71}$$

and

$$De = c^2\beta_S\left[(\beta_H-\beta_S)c^2\right]^{\frac{(D-1)}{2}}\Gamma\left(-\frac{D-1}{2}\right)+\frac{c^2\beta_S}{(m_0)^{\frac{(D-1)}{2}}\left(\frac{D-1}{2}\right)}$$
$$+\ \left[c^2(\beta_H-\beta_S)\right]^{\frac{(D+1)}{2}}\Gamma\left(-\frac{D+1}{2}\right)+\frac{1}{m_0^{\frac{(D+1)}{2}}\left(\frac{D+1}{2}\right)} \tag{72}$$

Therefore, from [Eq. (62)], [Eq. (71)] and [Eq. (72)] $\mathcal{A}_{\text{closed}}$ is given by

$$\left(\frac{N}{De}\right)_{\text{closed}} = \frac{\frac{c^2\beta_S}{(m_0)^{\frac{(D+3)}{2}}\left(\frac{D+3}{2}\right)}+\frac{1}{\left(\frac{D+5}{2}\right)(m_0)^{\frac{(D+5)}{2}}}}{\frac{c^2\beta_S}{m_0^{\frac{(D-1)}{2}}\left(\frac{D-1}{2}\right)}+\frac{1}{m_0^{\frac{(D+1)}{2}}\left(\frac{D+1}{2}\right)}} \tag{73}$$

and for $D=4$, we have for $\beta_H\to\beta_S$ ($M\to M_{\min}, r_S\to r_{\min}$)

$$\mathcal{A}_{\text{closed}} = \frac{M_{\min}^2 L_{PL}^6}{1260\pi m_0^2}\left(\frac{\frac{c^2\beta_S}{7}+\frac{1}{9m_0}}{\frac{c^2\beta_S}{3}+\frac{1}{5m_0}}\right) \tag{74}$$

From [Eq. (68)] and [Eq. (73)], we evaluate now the number $\mathcal{A}/r_S^6$ appearing in the expressions for r_+ [Eq. (53)], T_+ [Eq. (56)], M_+ [Eq. (60)], and τ_+ [Eq. (61)], for the two opposite limiting regimes $\beta_H\to\beta_S$ and $\beta_H\gg\beta_S$:

$$\left(\frac{\mathcal{A}_{\text{open}}}{r_{\min}^6}\right)_{\beta_H\to\beta_S} \simeq \frac{(\beta_H-\beta_S)}{\beta_S}\cdot\frac{16}{315}\left(\frac{\pi}{b}\right)^3\left(\frac{M_S}{M_{PL}}\right)^2\left(\frac{M_S}{m_0}\right)$$
$$\simeq (\beta_H-\beta_S)\cdot\frac{16}{315}\left(\frac{\pi}{b}\right)^3\left(\frac{M_S}{M_{PL}}\right)^2\frac{M_S^2c^2}{bm_0} \ll 1 \tag{75}$$

$$\left(\frac{\mathcal{A}_{\text{closed}}}{r_{\min}^6}\right)_{\beta_H\to\beta_S} \simeq \frac{16}{735b}\left(\frac{\pi}{b}\right)^3\left(\frac{M_S}{M_{PL}}\right)^2\left(\frac{M_S}{m_0}\right)^2 \ll 1 \tag{76}$$

In the opposite (semiclassical) regime $\beta_H\gg\beta_S$ i.e $T_H\ll T_S$, we have from [Eq. (66)], [Eq. (67)], [Eq. (69)] and [Eq. (70)]

$$\left(\frac{N}{De}\right)^{\text{open}}_{\beta_H\gg\beta_S} \simeq \frac{1}{m_0^2} \simeq \left(\frac{N}{De}\right)^{\text{closed}}_{\beta_H\gg\beta_S} \tag{77}$$

as

$$\beta_H m_0 c^2 = 8\pi \left(\frac{M}{M_{PL}}\right) \left(\frac{m_0}{M_{PL}}\right) \gg 1 \tag{78}$$

$(m_0, M \gg M_{PL})$. Then, from [Eq. (62)]

$$\left(\frac{\mathcal{A}}{r_S^6}\right)^{\text{open/closed}}_{\beta_H \gg \beta_S} \simeq \frac{1}{80640\pi} \left(\frac{M_{PL}}{M}\right)^4 \left(\frac{M_{PL}}{m_0}\right)^2 \ll 1 \tag{79}$$

That is, in this regime, we consistently recover $r_+ \simeq r_S$, $T_+ \simeq T_H$, $M_+ \simeq M$ and $\tau_+ \simeq \tau_H = \frac{k_B c}{6\sigma G\hbar T_H^3}$

6 CONCLUSIONS

We have suitably combined QFT and quantum string theory in the black hole background in the framework of the string analogue model (or thermodynamical approach).

We have computed the quantum string emission by a black hole and the back reaction effect on the black hole in the framework of this model. A clear and precise picture of the black hole evaporation emerges.

The QFT semiclassical regime and the quantum string regime of black holes have been identified and described.

The Hawking temperature T_H is the intrinsic black hole temperature in the QFT semiclassical regime. The intrinsic string temperature T_S is the black hole temperature in the quantum string regime. The two regimes are mapped one into another by the $\mathcal{R}$ - "Dual" transform.

String theory properly describes black hole evaporation: because of the emission, the semiclassical BH becomes a string state (the "minimal" BH), and the emitted string gas becomes a condensed microscopic state (the "minimal" BH) due to a phase transition. The last stage of the radiation in string theory, makes such a transition possible.

The phase transition undergone by the string gas at the critical temperature T_S represents (in the thermodynamical framework) the back reaction effect of the string emission on the BH.

The $\mathcal{R}$ - "Dual" relationship between QFT black holes and quantum strings revealed itself very interesting. It appears here this should be promoted to a Dynamical operation: evolution from classical to quantum (and conversely).

Cosmological evolution goes from a quantum string phase to a QFT and classical phase. Black hole evaporation goes from a QFT semiclassical phase to a string phase. The Hawking temperature, which we know as the black hole temperature, becomes the string temperature for the "string black hole".

Acknowledgements

M.R.M. acknowledges the Spanish Ministry of Education and Culture (D.G.E.S.) for partial financial support (under project : "Plan de movilidad personal docente e investigador") and the Observatoire de Paris - DEMIRM for the kind hospitality during this work.

Partial financial support from NATO Collaborative Grant CRG 974178 is also acknowledged. (N.S)

References

1. S. W. Hawking, Comm. Math. Phys. 43 (1975) 199.

2. H.J. de Vega, N. Sánchez, Nucl. Phys. B309 (1988) 522; B309 (1988) 577.

3. M. Ramon Medrano and N. Sánchez "*QFT, String Temperature and the String Phase of De Sitter Spacetime*", hep-th/9904015.

4. R. Hagedorn, Suppl. Nuovo Cimento 3, 147 (1965).

5. R.D. Carlitz, Phys. Rev. 5D (1972) 3231.

6. N. Sánchez, Phys. Rev. D18 (1978) 1030.

7. M.J. Bowick, L.S. Smolin, L.C. Wijewardhana, Phys. Rev. Lett. 56, 424 (1986).

8. K. Huang, S. Weinberg, Phys. Rev. Lett 25, 895 (1970).
M.B. Green, J.H. Schwarz, E. Witten, "*Superstring Theory*", Vol I. Cambridge University Press, 1987.

9. G.W. Gibbons in "*General Relativity, An Einstein Centenary Survey*", Eds. S.W. Hawking and W. Israel, Cambridge University Press, UK (1979).

10. N.D. Birrell, PCW. Davies "*Quantum Fields in Curved Space*", Cambridge University Press, UK (1982).

11. D.N. Page, Phys. Rev. D13 (1976) 198.

12. J.M. Bardeen, Phys. Rev. Lett. 46 (1981) 382.

13. V.P. Frolov and A.I. Zelnikov Phys. Lett B115 (1982) 372.

14. C.O. Lousto and N. Sánchez, Phys. Lett B212 (1988) 411 and Int. J. Mod. Phys. A4 (1989) 2317.

NONLOCALITY, ENTANGLEMENT AND QUANTUM COMPUTERS

ZHAO-LI ZHANG and ZHEN-JIU ZHANG
Center for Relativity studies, Department of Physics
Central China Normal University
Wuhan, 430079, China

ABSTRACT

In this paper we introduce the developments and develop our point of view in the field of the entanglement, nonlocality and quantum computation. It seems that the recently theoretical and experimental developments on entanglement and nonlocality are solid and important for our understanding the implication of quantum mechanics and that the developments on quantum computers are very important, but further developments will not be easy, therefore will be wonderful and attractive.

1. Introduction

"...trying to find a **computer simulation of physics**, seems to me to be an excellent program to follow out...and I'm not happy with all the analyses that go with just the classical theory, because **NATURE ISN'T CLASSICAL**, , ...and if you want to make a simulation of nature, you'd better **MAKE IT QUANTUM MECHANICAL**, and ... it's a wonderful problem because it doesn't look so easy." Richard Feynman[1].

Since the discussion between Einstein and Bohr which culminated in 1935 with the Einstein-Podolsky-Rosen[2] incompleteness argument and its rebuttal by Bohr[3], a great deal has been written and recently a lot of experiments has been performed. The famous Bell inequality[4] in 1964, the article published by Bell in 1966 and the famous experiment by Aspect[5] in 1981 and 1982 push us to recognize the nonlocality as an important quantum phenomenon, even though how to understand the underline idea is not clear for me. The entangled state becomes a very important subject for us. The preparation of entangled EPR-pairs and the quantum teleportation[6] have been realized in laboratory in 1997. The first entanglement of three photons and the first experimental test of quantum nonlocality[6] have realized in experiment in 1999.

A new area called quantum information combined of information theory, quantum mechanics and computer sciences has been established[7]. The applications of quantum information such as quantum communications and quantum computation[7] are also active areas.

The limitation both in principle and practical of now-a-days computers we are using, which are called classical computers, and the impossibility of classical computer simulation of quantum systems show the idea to develop quantum computers. The investigation shows that no principle difficulty has been found for quantum computers. The theoretical and experimental development on quantum computation seems optimistic since Shor developed his quantum algorithm[8] in 1994. The theoretical proposals and results of experimental realization of quantum computation[9] have appeared in 1998 and 1999. The advantages of quantum computation and its applications in cryptography and quantum teleportation attract a lot of interests of scientists and engineers.

The basic idea of quantum computers is entanglement and nonlocality of quantum mechanics. The realization of quantum computation is important both for entering a new quantum information age and for further understanding and development of quantum mechanics.

In this paper we introduce the developments and develop our point of view in the field of the entanglement, nonlocality and quantum computation. In the second section, we introduce the EPR pair, give the definition of entanglement, and develop the matrix expression and the density matrix expression for entangled states, pure state entanglement and three photons entanglement. Then introduce the experimental test of quantum nonlocality. In the third section, we discuss the quantum teleportation and physics behind it. In the fourth section, we concentrate on the quantum computation. We begin with the limitation of classical computers, give an introduction to quantum computation, quantum computers, quantum information, quantum bits, quantum gates, quantum registers and quantum parallelism, and then develop Shor algorithm and discrete Fourier transformation in details. We are regretful to omit the introduction and discussion on experimental developments on quantum computers. We do not think it is a good time to do so.

As a conclusion, we have confidence to think that the recently theoretical and experimental developments on entanglement and nonlocality are solid and important for our understanding quantum mechanics and that the developments on quantum computers are very important, but further developments will not be easy, therefore will be wonderful and attractive.

2. Entanglement and Nonlocality

2.1 EPR Pair

Einstein, Podolsky and Rosen (1935) first proposed the problem of nonlocality in a famous paper. The most discussion these days uses a simpler version due to Bohm (1951). He considered a particle whose decay produces two spin-parties whose total spin angular momentum is zero. These particles move away from each other in opposite directions, and two observers, A and B, subsequently measure the components of their spins along various directions. The constraint on the total spin is that if both observers agree to measure the spin along a particular direction x, and if A gets the result of $+h/2$, then B will necessarily get the result $-h/2$, and if A gets $-h/2$ then L will necessarily get $+h/2$ (where h is Planck constant). The results will be the same for the measurements along any other direction.

With any bipartite pure state, we may associate a positive integer, the Schumidt number, which is the number of terms in the Schumidt decomposition of $|\psi>$:

$$|\psi>=\Sigma_{ij} a_{ij} |i>\otimes|j>, (i,j=1,2), \qquad (2.1)$$

where $|i>\in H_1$, $|j>\in H_2$, H_1 and H_2 are Hilbert spaces, and $|\psi>\in H_1\otimes H_2$. We define: $|\psi>$ is **entangled** (or nonseparable) if its Schumidt number is greater than one; otherwise, it is **separable** (or unentangled). Thus, a separable bipartite pure state is a direct product of pure states in $H_1\otimes$ and H_2,

$$|\phi>=|i>\otimes|j>, (i,j=1,2); \qquad (2.2)$$

then the reduced density matrices $\rho_1 = |i><i|$ and $\rho_2 = |j><j|$ are pure. Any state that can not be expressed as such a direct product is entangled; then

$$\rho = |\psi><\psi|. \qquad (2.3)$$

When we use the expression of matrix, we have

$$|1\rangle = \begin{pmatrix} 0 \\ 1 \end{pmatrix}, |0\rangle = \begin{pmatrix} 1 \\ 0 \end{pmatrix}, \langle 1| = (0 \quad 1), \langle 0| = (1 \quad 0) \qquad (2.4)$$

where $|X>^* = <X|$, "$*$" stands for complex conjugate. Therefore the wave function is a spinor with two elements:

$$|\Psi\rangle = \begin{pmatrix} \alpha \\ \beta \end{pmatrix}, \langle \Psi | = (\alpha^*, \quad \beta^*) \qquad (2.5)$$

The corresponding density matrices is

$$\rho = \begin{pmatrix} \alpha\alpha^* & \alpha\beta^* \\ \beta\alpha^* & \beta\beta^* \end{pmatrix}. \qquad (2.6)$$

The trace of the density matrices is

$$\mathrm{tr}\rho = \alpha\alpha^* + \beta\beta^* = |\alpha|^2 + |\beta|^2 = 1. \qquad (2.7)$$

The density matrices ρ satisfies the operator equation:

$$i\hbar(\partial/\partial t)\rho = [H,\rho], \tag{2.8}$$

where H is the Hamiltonian for the system and the commutator $[H,\rho]$ is defined as:

$$[H,\rho] \equiv H\rho - \rho H. \tag{2.9}$$

It can be verified that the expectation value of observable operator can be expressed in the density matrices form as:

$$\langle A\rangle = \mathrm{tr}(\rho A) = \mathrm{tr}(A\rho). \tag{2.10}$$

The spin polarization vector $\boldsymbol{P}$ is defined as:

$$P_x = 2Re\ (\alpha^*\beta),\quad P_y = 2Im\ (\alpha^*\beta),\quad P_z = |\alpha|^2 - |\beta|^2, \tag{2.11}$$

which points to the direction of the spin. The relation between density matrices and spin polarization vector is

$$\rho = (1/2)(I + \boldsymbol{P}\cdot\sigma)\ ,\ \boldsymbol{P}\cdot\sigma|\Psi\rangle = |\Psi\rangle,\quad \mathbf{P}^2 = I. \tag{2.12}$$

where σ are the Pauli matrices:

$$\sigma_x = \begin{pmatrix} 0 & 1 \\ 1 & 0 \end{pmatrix}, \sigma_y = \begin{pmatrix} 0 & -i \\ i & 0 \end{pmatrix}, \sigma_z = \begin{pmatrix} 1 & 0 \\ 0 & -1 \end{pmatrix}. \tag{2.13}$$

And we have:

$$\mathrm{tr}\sigma = 0,\quad \boldsymbol{P} = \mathrm{tr}(\rho\sigma). \tag{2.14}$$

In quantum theory, suppose first that the measurements are made along the z-axes of the two observers. The spin part of the state of the two particles can be written in terms of the associated eigenvectors as

$$|\psi\rangle = (1/\sqrt{2})(|\uparrow\rangle|\downarrow\rangle - |\downarrow\rangle|\uparrow\rangle), \tag{2.15}$$

where, for example, $|\uparrow\rangle|\downarrow\rangle$ is the state in which particles 1 and 2 have spin $+h/2$ and $-h/2$ respectively. The expression (2.15) is one of the four mutually orthogonal maximally entangled states of two particles. The interpretation of the entangled state $|\psi>$ is straightforward in a series of repeated measurements by A, a selection is made of the pairs of particles for which the measurement of her particle gave spin-up, then —with probability one —a series of measurements by B on her particle in these pairs will yield spin-down. Similarly, if A finds spin-down then, with probability one, B will find spin-up.

But difficulties arise if one tries to enforce a more realist interpretation of the entangled state. The obvious question is how the information about each observer's individual results "gets to" the other particle to guarantee that the result obtained by the second observer will be the correct one.

One might be tempted to invoke the reasoning of classical physics and argue that both particles possess appropriate value all the time. However, the only way in standard quantum theory of guaranteeing that a certain result will be obtained is if the state is an eigenvector of the observable concerned. But the state $|\psi>$ is not of this type: rather it displays the typical features of quantum entanglement and is in fact a superposition of such states. There is also a question of whether this picture is compatible with special relativity. If the measurements by the two observers are space-like separated (which could easily be arranged) then which of them makes the first measurement, and hence in the standard interpretation causes the state-vector reduction of the second one.

Einstein, Podolsky and Rosen considered these issues, and concluded that the difficulties could be resolved in one of only two ways:

1. When A makes her measurement, the result communicates itself at once in some way to particle 2, and converts its state into the appropriate eigenvector. Or,
2. Quantum theory is incomplete and provides only a partial specification of the actual state of the system.

In contemplating the first possibility it must be appreciated that the two particles may have moved a vast distance apart before the first measurement is made, and therefore any "at once" mode of communication would be in violent contradiction with the spirit (if not the law) of special relativity. It is not surprising that Einstein was not very keen on this alternative! An additional objection involves the lack within quantum theory itself of any idea about how this nonlocal effect is supposed to take place, so in this sense the theory would be incomplete anyway. Einstein, Podolsky and Rosen came to the conclusion that the theory is indeed incomplete, although they left open the question of the correct way in which to "complete" it.

This means that if the predictions of quantum theory are experimentally valid in this region then any idea of systems possessing and individual values for observables must necessarily involves an essential nonlocality.

A number of experiments have been performed over the last 20 years, the most famous of which is the work of Alain Aspect and collaborators who checked the correlation between photons in a variety of configurations (Aspect, Graingier and Roger 1981.1982).

In any case, it seems that we have to accept the existence of a strange **nonlocality**.

Entangled states arise, naturally, as a result of interactions between particles, such as when a pair of particles are created simultaneously under the requirement that some attribute, such as spin or polarization, be conserved. According to quantum theory, although the exact state of each particle is not determined until a measurement is made, their states must, nonetheless, be correlated. Thus whenever a measurement is made on one member of the entangled pair, the states of both particles become definite and the entanglement ceases. The entangled pair is called **EPR pair**.

Hoi-Kwong Lo (1999) et al. construct an explicit procedure which demonstrates that for bipartite pure states, in the asymptotic limit, entanglement can be concentrated or diluted with vanishing classical communication cost. Entanglement of bipartite pure states is thus established as a truly interconvertible resource.

Quiroga and Johnson (1999) show that excitons in coupled quantum dots are ideal candidates for reliable preparation of entangled states in solid-state systems. An optically controlled exciton transfer process is shown to lead to the generation of Bell and Greenberger-Horne-Zeilinger states in systems comprising two and three coupled dots, respectively. The strength and duration of selective light pulses for producing maximally entangled states are identified by both analytic and full numerical solution of the quantum dynamical equations.

2.2 *Entanglement of Pure State*

Vidal (1999) has presented an optimal local conversion strategy between any two pure states of a bipartite system. It is optimal in that the probability of success is the largest achievable if the parties which share the system, and which can communicate classically, are only allowed to act locally on it. The study of optimal local conversions sheds some light on the entanglement of a single copy of a pure state. They propose a quantification of such an entanglement by means of a finite minimal set of new measures from which the optimal probability of conversion follows.

For the bipartite system with two pure states, there are four maximal entangled states, which are mutual orthogonal. They are

$$|\Phi^+\rangle=(1/\sqrt{2})(|00\rangle+|11\rangle),\ |\Phi^-\rangle=(1/\sqrt{2})(|00\rangle-|11\rangle); \tag{2.16}$$
$$|\psi^+\rangle=(1/\sqrt{2})(|01\rangle+|10\rangle),\ |\psi^-\rangle=(1/\sqrt{2})(|01\rangle-|10\rangle). \tag{2.17}$$

One can form a representation of entanglement states based on those four states as basics. The maximal entangled states mean that the density matrices of the entangled states possess the following properties: for the entangled state

$$|\Phi^+\rangle=(1/\sqrt{2})(|00\rangle+|11\rangle), \tag{2.18}$$

after getting the trace for the second qubit, the density matrices of the first qubit will be

$$\rho_1=tr_2(|\Phi^+\rangle\langle\Phi^+|)=(1/2)I_1, \tag{2.19}$$

where I_1 is the identical matrices; Similarly, after getting the trace for the first qubit, the density matrices of the second qubit will be

$$\rho_2 = tr_1(|\Phi^+\rangle\langle\Phi^+|) = (1/2)I_2, \tag{2.20}$$

where I_2 is the identical matrices too.

By using a local unitary transformation

$$\sigma_3 = \begin{pmatrix} 1 & 0 \\ 0 & -1 \end{pmatrix},$$

one can get the changes: $|0\rangle \Leftrightarrow |1\rangle$, $|\Phi^+\rangle \Leftrightarrow |\Phi^-\rangle$, $|\psi^+\rangle \Leftrightarrow |\psi^-\rangle$; By using a local unitary transformation

$$\sigma_1 = \begin{pmatrix} 0 & 1 \\ 1 & 0 \end{pmatrix},$$

one can get the changes: $|\Phi^+\rangle \Leftrightarrow |\psi^+\rangle$, $|\Phi^-\rangle \Leftrightarrow -|\psi^-\rangle$. Local unitary transformation does not change the density matrices of individual qubit, but the information that they treated is unreadable.

Jonathan and Plenio (1999) find a minimal set of necessary and sufficient conditions for the existence of a local procedure that converts a finite pure state into one of a set of possible final states. This result provides a powerful method for obtaining optimal local entanglement manipulation protocols for pure initial states. As an example, they determine analytically the optimal disyllable entanglement for arbitrary finite pure states. They also construct an explicit protocol achieving this bound.

2.3 The First Entanglement of Three Photons

The first entanglement of three photons has been experimentally demonstrated by researchers at the University of Innsbruck (Bouwmeester et. al.1999). Individually, an entangled particle has properties (such as momentum) that are indeterminate and undefined until the particle is measured or otherwise disturbed. Measuring one entangled particle, however, defines its properties and seems to influence the properties of its partner or partners instantaneously, even if they are light years apart. In the present experiment, sending individual photons through a special crystal sometimes converted a photon into two pairs of entangled photons. After detecting a "trigger" photon, and interfering two of the three others in a beamsplitter, it became impossible to determine which photon came from which entangled pair. As a result, the respective properties of the three remaining photons were indeterminate, which is one way of saying that they were entangled (the first such observation for three physically separated particles). The researchers deduced that this entangled state is the long-coveted GHZ state proposed by physicists Daniel Greenberger, Michael Horne, and Anton Zeilinger in the late 1980s. In addition to facilitating more advanced forms of quantum cryptography, the GHZ state will help provide a nonstatistical test of the foundations of quantum mechanics. Albert Einstein, troubled by some implications of quantum science, believed that any rational description of nature is incomplete unless it is both a local and realistic theory: "realism" refers to the idea that a particle has properties that exist even before they are measured, and "locality" means that measuring one particle cannot affect the properties of another, physically separated particle faster than the speed of light. But quantum mechanics states that realism, locality--or both--must be violated. Previous experiments have provided highly convincing evidence against local realism, but these "Bell's inequalities" tests require the measurement of many pairs of entangled photons to build up a body of statistical evidence against the idea. In contrast, studying a single set of properties in the GHZ particles could verify the predictions of quantum mechanics while contradicting those of local realism.

Brassard et. al. (1999) investigate the amount of communication that must augment classical local hidden variable models in order to simulate the behavior of entangled quantum systems. They consider the scenario where a bipartite measurement is given from a set of possibilities and the goal is to obtain exactly the same correlations that arise when the actual quantum system is measured. They show that, in the case of a single pair of qubits in a Bell state, a constant number of bits of communication is always sufficient—regardless of the number of measurements under consideration. They also show that, in the case of a system of n Bell states, a constant times 2n bits of communication is necessary.

Mor (1999) consider a disentanglement process which transforms a state of two subsystems into an unentangled (i.e., separable) state, while not affecting the reduced density matrix of either subsystem. Recently, Terno (1999) showed that a physically allowable process cannot disentangle a tensor product of its reduced density matrices into an arbitrary state. They show that there are sets of states which can be disentangled, but only into separable states other than the product of the reduced density matrices, and other sets of states which cannot be disentangled at all. Thus, they prove that a universal disentangling machine cannot exist.

2.4 The First Experimental Test of Quantum Nonlocality

Pan et al. (1999) report the first experimental test of quantum nonlocality in three-photon GHZ entanglement. This test addresses a strong conflict with local realism for those cases where quantum theory makes definite, i.e. nonstatistical, predictions. This is in contrast to the case of EPR Pair experiments, where the conflict with local realism only arises for the statistical predictions.

They report the first experimental test of local realism versus quantum mechanics that utilizes three-particle entanglement. This test, put forward by Greenberger, Horne and Zeilinger (GHZ), addresses a strong conflict with local realism for those cases where quantum theory makes definite, i.e. nonstatistical, predictions. This is in contrast to the case of Einstein-Podolsky-Rosen experiments with two entangled particles, where according to Bell's theorem the conflict with local realism only arises for the statistical predictions. The experimental results are in agreement with the quantum prediction and in distinct conflict with a local realistic interpretation. Ever since its introduction by Schrödinger entanglement has commanded a central position in the discussions of the interpretation of quantum mechanics. Originally that discussion has focused on the proposal by Einstein, Podolsky and Rosen (EPR) of measurements performed on two spatially separated entangled particles. Most significantly John Bell then showed that there is a conflict between any attempt to explain the correlations observed in such systems by a local realistic model and the predictions made by quantum mechanics. In the derivation of Bell's inequalities one makes the seemingly innocuous assumption that perfect correlations can be understood using such a local realistic model and the conflict then arises for the statistical predictions of quantum theory.

An increasing number of experiments on entangled particle pairs having confirmed the statistical predictions of quantum mechanics (Weihs et al., 1998) have thus provided increasing evidence against local realistic theories. Yet, one might find some comfort in the fact that such a realistic and thus classical picture can explain perfect correlations and is only in conflict with statistical predictions of the theory. After all, quantum mechanics is statistical in its core structure. In other words, for entangled particle pairs a local realistic model can explain the cases where the result of a measurement on one particle can definitely be predicted on the basis of a measurement result on the other particle. It is only that subset of statistical correlations where the measurement results on one particle can only be predicted with a certain probability, which cannot be explained by such a model. Yet in 1989 Greenberger, Horne and Zeilinger (GHZ) showed it that for certain three- and four-particle states (Greenberger et al., 1989,1990) a conflict with local realism arises even for perfect correlations. That is, even for those cases where, based on measurement on N- 1 of the particles, the result of the measurement on particle N can be predicted with certainty. Local realism and quantum mechanics here both make definite but completely opposite predictions. A particularly elegant demonstration of that conflict is due to Mermin (1990).

Utilizing a recently developed source for three-photon GHZ-entanglements, this work presents a first realization of such a three-particle test against local realism.

3 Quantum Teleportation

3.1 Quantum Teleportation

Teleportation is the name given by science fiction writers to the feat of making an object or person disintegrate in one place while a perfect replica appears somewhere else. How this is accomplished is usually not explained in detail, but the general idea seems to be that the original object is scanned in such a way as to extract all the information from it, then this information is transmitted to the receiving location and used to construct the replica, not necessarily from the actual material of the original, but perhaps from atoms of the same kinds, arranged in exactly the same pattern as the original. A teleportation machine would be like a fax machine, except that it would work on 3-dimensional objects as well as documents, it would produce an exact copy rather than an approximate facsimile, and it would destroy the original in the process of scanning it. A few science fiction writers consider teleporters that preserve the original, and the plot gets complicated when the original and teleported versions of the same person meet; but the more common kind of teleporter destroys the original, functioning as a super transportation device, not as a perfect replicator of souls and bodies.

Two years ago an international group of six scientists, including IBM Fellow Charles H. Bennett, confirmed the intuitions of the majority of science fiction writers by showing that perfect teleportation is indeed possible in principle, but only if the original is destroyed. Meanwhile, other scientists are planning experiments to demonstrate teleportation in microscopic objects, such as single atoms or photons, in the next few years. But science fiction fans will be disappointed to learn that no one expects to be able to teleport people or other macroscopic objects in the foreseeable future, for a variety of engineering reasons, even though it would not violate any fundamental law to do so.

Until recently, teleportation was not taken seriously by scientists, because it was thought to violate the uncertainty principle of quantum mechanics, which forbids any measuring or scanning process from extracting all the information in an atom or other object. According to the uncertainty principle, the more accurately an object is scanned, the more it is disturbed by the scanning process, until one reaches a point where the object's original state has been completely disrupted, still without having extracted enough information to make a perfect replica. This sounds like a solid argument against teleportation: if one cannot extract enough information from an object to make a perfect copy, it would seem that a perfect copy couldn't be made. But the six scientists found a way to make an end-run around this logic, using a celebrated and paradoxical feature of quantum mechanics known as the Einstein-Podolsky-Rosen effect. In brief, they found a way to scan out part of the information from an object A, which one wishes to teleport, while causing the remaining, unscanned, part of the information to pass, via the Einstein-Podolsky-Rosen effect, into another object C which has never been in contact with *A*. Later, by applying to C a treatment depending on the scanned-out information, it is possible to maneuver C into exactly the same state as A was in before it was scanned. *A* itself is no longer in that state, having been thoroughly disrupted by the scanning, so what has been achieved is teleportation, not replication.

The unscanned part of the information is conveyed from A to C by an intermediary object B, which interacts first with C and then with A. Can it really be correct to say "first with C and then with A"? Surely, in order to convey something from A to C, the delivery vehicle must visit A before C, not the other way around. But there is a subtle, unscannable kind of information that, unlike any material cargo, and even unlike ordinary information, can indeed be delivered in such a backward fashion. This subtle kind of information, also called "Einstein-Podolsky-Rosen (EPR) correlation" or "entanglement", has been at least partly understood since the 1930s when it was discussed in a famous paper by Albert Einstein, Boris Podolsky, and Nathan Rosen. In the 1960s John Bell showed that a pair of entangled particles, which were once in contact but later move too far apart to interact directly, can exhibit individually random behavior that is too strongly correlated to be explained by classical statistics. Experiments on photons and other particles have repeatedly confirmed these correlations, thereby providing strong evidence for the validity of quantum mechanics, which neatly explains them. Another well-known fact about EPR correlations is that they cannot by themselves deliver a meaningful and controllable message. It was thought that their only usefulness was in proving the validity of quantum mechanics. But now it is known that, through the phenomenon of quantum teleportation, they can deliver exactly that part of the information in an object, which is too delicate to be scanned out, and delivered by conventional methods.

In conventional facsimile transmission the original is scanned, extracting partial information about it, but remains more or less intact after the scanning process. The scanned information is sent to the receiving station, where it is imprinted on some raw material (e.g. paper) to produce an approximate copy of the original. In quantum teleportation two objects B and C are first brought into contact and then separated. Object B is taken to the sending station, while object C is taken to the receiving station. At the sending station, object B is scanned together with the original object A which one wishes to teleport, yielding some information and totally disrupting the state of A and B. The scanned information is sent to the receiving station, where it is used to select one of several treatments to be applied to object C, thereby putting C into an exact replica of the former state of A.

3.2 Physics behind Quantum Teleportation

Teleportation, as it is meant here, relies upon a typical quantum phenomenon known as the EPR effect, named after its discoverers Albert Einstein, Boris Podolsky, and Nathan Rosen (Einstein et al. 1935). The EPR effect describes the enduring interconnection between pairs of quantum systems that had, at some earlier time, interacted with one another. Surprisingly, many interpretations of quantum theory predict that such an interaction is instantaneous, unmediated, and immune to the nature of the intervening medium.

It certainly did not sit well with Einstein, Podolsky, and Rosen who tried to use the apparent absurdity of the prediction to prove that quantum mechanics gave an incomplete description of physical reality. As quantum teleportation relies crucially on such an "action at a distance" effect, it is important to take a minor diversion to convince you that the effect is real and that reality is, in fact, nonlocal.

The scientific evidence that arose in the mid-1980s proves, empirically, that reality is nonlocal and that is exploited in quantum teleportation.

Hoi Fung Chau and H.-K. Lo (1996) show that the entropy of entanglement of a state characterizes its ability to teleport. In particular, in order to teleport faithfully an unknown quantum N-state, the two users must share an entangled state with at least $\log_2 N$ bits entropy of entanglement. Their result, therefore, provides an alternative interpretation for entanglement purification.

Teleporting the quantum state of an object A to object C requires a third object B. B and C are created together as a quantum connected "EPR" pair. B then meets A and transmits information about A to C. In addition, partial information about A is transmitted to C by an ordinary signal. The net teleportation effect is, therefore, a combination of ordinary retarded classical signal transmission forward in time together with an extraordinary advanced quantum transmission of information backward in time.

4. Quantum Computation

4.1 The Limitation of Classical Computers

Civilization has advanced as people discovered new ways of exploiting various physical resources such as materials, *energies* and *information*.

Today's advanced lithographic techniques can squeeze fraction of micron wide logic gates and wires onto the surface of silicon chips.

Computer chips continue to shrink. But the discovery of silicon dioxide must be at least four to five atoms thick to function as an insulator suggests that silicon-based microchips will reach the limits of miniaturization early next century.

Soon they will inevitably reach a point where logic gates are so small that they are made out of only a handful of atoms. On the atomic scale matter obeys the rules of quantum mechanics.

A classical algorithm to factor an integer number N on a classical computer is based on checking the remainder of the division of N by some number p smaller than square root of N. If the remainder is 0, we conclude that p is a factor. This method is in fact very inefficient: with a computer that can test for 10^{10} different p's per second (this is faster than any computer ever built), the average time to find the factor of a 60-digit long number would exceed the age of the universe!

The story of quantum computation started as early as 1982, when the physicist Richard Feynman considered simulation of quantum-mechanical objects by other quantum systems. However, the unusual power of quantum computation was not really anticipated until the 1985 when David Deutsch of the University of Oxford published a crucial theoretical paper in which he described a universal quantum computer. After the Deutsch paper, one found something interesting for quantum computers to do. At the time all that could be found were a few rather contrived mathematical problems and the whole issue of quantum computation seemed little more than an academic curiosity. It all changed rather suddenly in 1994 when Peter Shor from AT&T's Bell Laboratories in New Jersey devised the first quantum algorithm that, in principle, can perform efficient factorisation. This is one thing very useful that only a quantum computer could do. Difficulty of factorisation underpins security of many common methods of encryption; for example, RSA the most popular public key cryptosystem, which is often used to protect electronic bank accounts security from the difficulty of factoring large numbers.

4.2 Introduction to Quantum Computation

In the early 1980s Richard Feynman (among others) began to investigate the generalization of conventional information science concepts to quantum physical processes. He considered the representation of binary numbers by orthogonal quantum states (|0> or |1>) of two-level quantum systems. (The representation of a single bit of information in this form has come to be known as a "qubit.") With two or more qubits it becomes possible to consider quantum logical "gate" operations in which a controlled interaction between qubits produces a (coherent) change in the state of one qubit that is contingent upon the state of another. These gate operations are the building blocks of a quantum computer (QC), which in principle is a very much more powerful device than any classical computer because the superposition principle allows an extraordinarily large number of computations to be performed simultaneously. In 1994 it was shown that this "quantum parallelism" could be used to efficiently find the prime factors of composite integers. Integer factorization and related problems that are computationally intractable with conventional computers are the basis for the security of modern public key cryptosystems. However, a quantum computer running at desktop PC speeds could break the "keys" of these cryptosystems in only seconds (as opposed to the months or years required with conventional computers). This single result has turned quantum computation from a strictly academic exercise into a subject whose practical feasibility must be urgently determined.

4.3 Quantum Computers

The architecture of a quantum computer is conceptually very similar to a conventional computer: multi-(qu)bit registers are used to input data; the contents of the registers undergo logical "gate" operations to effect the desired computation under the control of an algorithm; and finally a result must be read out as the contents of a register. The principal obstacles to constructing a practical quantum computer are: the difficulty of engineering the quantum states required; the phenomenon of "decoherence," which is the propensity for these quantum states to lose their coherence properties through interactions with the environment; and the quantum measurements required to read-out the result of a quantum computation. The first proposals for practical quantum computation hardware, based on various exotic technologies, suffered from one or more of these problems. However, in 1994 it was proposed that the basic logical "gate" operations of quantum computation could be experimentally implemented with laser manipulations of cold, trapped ions: a qubit would comprise the ground (S) state (|0) and a suitably chosen metastable excited (D) state (|1) of an ion isolated from the environment by the electromagnetic fields of a linear radiofrequency quadrupole (RFQ) ion trap. The principal components of this technology are already well developed for frequency standard and high-precision spectroscopy work. Existing experimental data suggest that adequate coherence times may be achievable and a read-out method based on so-called "quantum jumps" has already been demonstrated with single trapped ions. Furthermore, the NIST group in Boulder has recently demonstrated a quantum logical gate operation with a version of this idea. They are developing an ion trap quantum computer experiment using calcium ions with the ultimate objective of performing multiple gate operations on a register of several qubits (and possibly small computations) in order to determine the potential and physical limitations of this technology.

In principle we know how to build a quantum computer; we can start with simple quantum logic gates and try to integrate them together into quantum circuits. A quantum logic gate, like a classical gate, is a very simple computing device that performs one elementary quantum operation, usually on two qubits, in a given period of time. Of course, quantum logic gates are different from their classical counterparts because they can create and perform operations on quantum superposition. However if we keep on putting quantum gates together into circuits we will quickly run into some serious practical problems. The more interacting qubits are involved the harder it tends to be to engineer the interaction that would display the quantum interference. Apart from the technical difficulties of working at single-atom and single-photon scales, one of the most important problems is that of preventing the surrounding environment from being affected by the interactions that generate quantum superposition. The more components the more likely it is that quantum computation will spread outside the computational unit and will irreversibly dissipate useful information to the environment. This process is called decoherence. Thus the race is to engineer sub-microscopic systems in which qubits interact only with themselves but not with the environment.

Some physicists are pessimistic about the prospects of substantial experimental advances in the field.

Can we then control nature at the level of single photons and atoms? Yes, to some degree we can! For example in the so called cavity quantum electrodynamics experiments, which were performed by Serge Haroche, Jean-Michel Raimond and colleagues at the Ecole Normale Superieure in Paris, atoms can be controlled by single photons trapped in small superconducting cavities. Another approach advocated by Christopher Monroe, David Wineland and coworkers in Boulder, USA, and uses ions sitting in a radio-frequency trap. Ions interact with each other exchanging vibrational excitations and a properly focused and polarised laser beam can separately control each ion.

Experimental and theoretical research in quantum computation is accelerating worldwide. New technologies for realizing quantum computers are being proposed, and new types of quantum computation with various advantages over classical computation are continually being discovered and analyzed and we believe some of them will bear technological fruit. From a fundamental standpoint, however, it does not matter how useful quantum computation turns out to be, nor does it matter whether we build the first quantum computer tomorrow, next year or centuries from now. The quantum theory of computation must in any case be an integral part of the worldview of anyone who seeks a fundamental understanding of the quantum theory and the processing of information.

4.4 Quantum Information

Quantum information is an exciting new paradigm for information science, which makes use of the counterintuitive concept of quantum superposition of information. This new concept raises the possibility of capabilities for information transmission, storage, and manipulation that are simply impossible with conventional information technologies. In the past few years there have been advances in the experimental study of the foundations of quantum mechanics, photonics, and atomic physics that have made accessible these novel uses of quantum states. Moreover, important applications of quantum mechanical concepts to information security and information assurance have been identified, and so this field has recently undergone a dramatic revolution from an essentially academic subject to one with an enormous potential to revolutionize computer science. The realization of these new information science concepts requires the ability to "engineer" quantum mechanical (coherent) states of several particles, which have hitherto only been used in quite limited forms for testing the foundations of quantum mechanics.

The foundations of information science were laid out during and shortly after World War II and tacitly assumed that information, whether in the form of ink on paper or voltages in a microprocessor, would be represented by processes obeying classical physics. However, in the early 1980s Richard Feynman and Charles Bennett (among others) began to investigate the generalization to information represented by quantum physical processes. That is, they considered the representation of binary numbers by orthogonal quantum states ($|0\rangle$ or $|1\rangle$) of some suitable two-level quantum system. (The representation of a single bit of information in this form has come to be known as a "qubit.") Examples of physical systems that permit such a qubit representation are biquitous: vertical and horizontal photon polarization states; single-photon interference states in which a photon can emerge from one or the other exit ports of a Mach-Zehnder interferometer; and the electronic ground and first (metastable) excited state of a (trapped) ion, to list only three. From this pioneering work it has been shown that quantum mechanics opens up powerful new methods for information storage, transmission and manipulation because of the superposition principle, the indivisibility of quanta and the peculiarities of measurement in quantum mechanics.

4.5 Quantum bits (qubits)

The quantum computers based on quantum mechanics can offer much more than cramming more and more bits to silicon and multiplying the clock-speed of microprocessors. It can support entirely new kind of computation with qualitatively new algorithms based on quantum principles!

From a physical point of view a **bit** is a physical system which can be prepared in one of the two different states representing two logical values: 0 or 1.

However, if we choose an atom as a physical system which can be prepared in, for example, exited state (representing 1) or fundamental state (representing 0), then quantum mechanics tells us that apart from the two distinct states the atom can be also prepared in a *coherent superposition* of the two states. This means that the atom is *both* in state 0 and state 1. It can be written as

$$|\psi\rangle = a\,|0\rangle + b\,|1\rangle\,, \tag{4.1}$$

where a and b are complex numbers. $|a|^2$and $|b|^2$ are probabilities of the states in $|0\rangle$ or $|1\rangle$ respectively.

An atom prepared in a superposition of two different states, and in general a quantum two state system, called a quantum bit or a **qubit**. Thus one qubit can encode at a given moment of time both 0 and 1.

Qubit = "quantum" + "bit"

A qubit is exactly what it sounds like: a bit of information represented by a quantum object, such as a single atom, ion, or photon. Just as a bit is the basic unit of information in a classical information system, a qubit is the basic unit of information of a quantum information system. The power of a qubit is that it is not limited to a value of 0 or 1. It can actually be in a "superposition" state of any combination of 0 or 1. This is the difference between a classical and a quantum computer, which theoretically allows quantum computers to be much more efficient at solving certain types of problems.

4.6 Quantum Logic gate

Any quantum system with at least two discrete and sufficiently separated energy levels is an appropriate candidate for a qubit. So far the most popular qubits among the experimentalists have been atoms of rubidium (ENS experiments) or beryllium (NIST experiments). In an atom, the energy levels of the various electrons are discrete. Two of them can be selected and labeled as logical 0 and 1. These levels correspond to specific excitation states of the electrons in the atom.

To see how it is possible to control the two logical values, let us simplify the situation and consider an idealized atom with a single electron and two energy levels, a ground state (identified with the value 0) and an excited state (value 1),

Let us imagine that the atom is initially in state 0 and that we want to effect the NOT operation. The NOT operation just negates the logical value so that 0 becomes 1 and 1 becomes 0. This is easily performed with atoms. By shining a pulse of light of appropriate intensity, duration and wavelength (the wavelength should match the energy difference between the two energy levels), it is possible to force the electron to change energy level. In our case, the electron initially in the ground state 0 will absorb some energy from the light pulse and transfer itself to the state 1. If the electron were initially in the state 1, the light pulse would cause transfer of the electron to state 0.

Logical NOT, described above, can be understood in completely classical terms, however, we can also perform purely quantum logical operations without any classical analogs. For instance, by using a light pulse of half the duration as the one needed to perform the NOT operation, we effect a half--flip between the two logical states. What does this mean? In substance, it means that operation 0 --->1 is only half performed and that the state of the atom after the half pulse is neither 0 or 1 but a coherent quantum superposition of both states. Just like the photon that travels through both A and B routes, the electron of the atom is in both 0 and 1 state. Acting on an atom in that state with another half pulse of light will complete the NOT operation, and set it in the 1 state. Because operation NOT is obtained as the result of two consecutive quantum "half--flips" this purely quantum operation is often called the square root of NOT. Using this kind of "half--pulses", it is possible to create superposition of states in the memory of a quantum computers (a memory would, of course, consist of many atoms), opening thus new ways to perform computations. Subsequent logical operations usually involve more complicated physical operations on more than one qubit.

Three quantum gates and the identical gate can be expressed by matrices

$$\vee=\begin{pmatrix}0&1\\1&1\end{pmatrix},\wedge=\begin{pmatrix}0&0\\0&1\end{pmatrix},\neg=\begin{pmatrix}0&1\\1&0\end{pmatrix},I=\begin{pmatrix}1&0\\0&1\end{pmatrix} \tag{4.2}$$

It is easy to verify that:

Identical gate I: $I|0\rangle=|0\rangle$, $I|1\rangle=|1\rangle$;

NOT gate $\neg$: $\neg|0\rangle=|1\rangle$, $\neg|1\rangle=|0\rangle$;

OR gate $\vee$: $\langle 1|\vee|1\rangle=1$, $\langle 1|\vee|0\rangle=1$, $\langle 0|\vee|1\rangle=1$, $\langle 0|\vee|0\rangle=0$.

AND gate $\wedge$: $\langle 1|\wedge|1\rangle=1$, $\langle 1|\wedge|0\rangle=0$, $\langle 0|\wedge|1\rangle=0$, $\langle 0|\wedge|0\rangle=0$.

It is also easy to verify that the relation between the expression of quantum logic gates and the Pauli matrices can be written as following:

$$\begin{gathered}
\vee=\sigma_-\sigma_+ +\sigma_- +\sigma_+, \qquad \wedge=\sigma_-\sigma_+, \\
\neg=\sigma_- +\sigma_+, \qquad I=\sigma_-\sigma_+ +\sigma_+\sigma_-, \\
s\equiv\sigma_+=\frac{1}{2}(\sigma_x+i\sigma_y)=\begin{pmatrix}0&1\\0&0\end{pmatrix}, \\
x\equiv\sigma_-=\frac{1}{2}(\sigma_x-i\sigma_y)=\begin{pmatrix}0&0\\1&0\end{pmatrix}, \\
\sigma_x=\begin{pmatrix}0&1\\1&0\end{pmatrix},\sigma_y=\begin{pmatrix}0&-i\\i&0\end{pmatrix}.
\end{gathered} \tag{4.3}$$

We can see that s-matrices stands for turn-up and x-matrices stands for turn-down:

$$s|1\rangle=|0\rangle,\quad x|0\rangle=|1\rangle. \tag{4.4}$$

The NOT gate

$$\neg = s + x. \tag{4.5}$$

An important quantum gate in quantum computers is the control-not gate or simplified as CN. It is a unitary and Hermit 2-digit quantum gate:

$$CN = |00\rangle\langle 00| + |01\rangle\langle 01| + |10\rangle\langle 11| + |11\rangle\langle 10|, \tag{4.6}$$

where the first bit is the control bit and the second one is called object bit. The function of CN is: when the control-bit is $|0\rangle$, the object-bit keeps unchanged (accomplished by the first two terms); when the control-bit is $|1\rangle$, the object-bit changes (accomplished by the last two terms). In the expression of matrices, CN-gate is :

$$CN = \begin{pmatrix} 1 & 0 & 0 & 0 \\ 0 & 1 & 0 & 0 \\ 0 & 0 & 0 & 1 \\ 0 & 0 & 1 & 0 \end{pmatrix}. \tag{4.7}$$

Another important quantum gate in quantum computers is the 3-digit F-gate. It is also a unitary and Hermit quantum gate:

$$\begin{aligned} F = {} & |000\rangle\langle 000| + |001\rangle\langle 001| + |010\rangle\langle 010| + |011\rangle\langle 011| \\ & + |100\rangle\langle 100| + |101\rangle\langle 110| + |110\rangle\langle 101| + |111\rangle\langle 111|. \end{aligned} \tag{4.8}$$

The left bit is control-bit and the following two bits are object-bit. The function of F-gate is: when the control-bit is $|0\rangle$, the object-bit keeps unchanged (accomplished by the first four terms); when the control-bit is $|1\rangle$, the object-bit changes (accomplished by the last four terms). For example,

$$F|001\rangle = |001\rangle; \; F|101\rangle = |110\rangle. \tag{4.9}$$

In occasions, we use creation operator a^+ and annihilation operator a:

$$a^+ = |1\rangle\langle 0|, \; a = |0\rangle\langle 1|, \tag{4.10}$$

then F-gate can be expressed as :

$$F = E + a^+a(b^+c + bc^+ - b^+b - c^+c + 2b^+bc^+c). \tag{4.11}$$

At last we consider the control-control-NOT gate or simplified as CCN-gate:

$$\mathrm{CCN}=|000\rangle\langle 000|+|001\rangle\langle 001|+|010\rangle\langle 010|+|011\rangle\langle 011| \\ +|100\rangle\langle 100|+|101\rangle\langle 101|+|110\rangle\langle 111|+|111\rangle\langle 110|, \qquad (4.12)$$

The left two bits are control-bit and the following bit is object-bit. The function of CCN-gate is: when the two control-bit is $|1\rangle$, the object-bit changes (accomplished by the last four terms).

4.7 Quantum Register

A quantum register is a set of qubits. For example, 6 is 110 in binary system, but it is $|1\rangle\otimes|1\rangle\otimes|0\rangle$ in a quantum register. Simply, let

$$|a\rangle \equiv |a_{n-1}\rangle\otimes|a_{n-2}\rangle......|a_1\rangle\otimes|a_0\rangle, \qquad (4.13)$$

express the value

$$a = 2^0 a_0 + 2^1 a_1 + ...+2^{n-1} a_{n-1}. \qquad (4.14)$$

For $|a\rangle$ and $|b\rangle (a\neq b)$, if we have

$$\langle a\,|\,b\rangle = \langle a_0\,|\,b_0\rangle\langle a_1\,|\,b_1\rangle\cdots\langle a_{n-1}\,|\,b_{n-1}\rangle = 0, \qquad (4.15)$$

$|a\rangle$ and $|b\rangle$ are said to be orthogonal. For an n-digit quantum register, the most general expression of a state is

$$|\Psi\rangle = \sum_{X=0}^{2^n-1} c_X |X\rangle. \qquad (4.16)$$

It is important that in the classical case for a 2-digit register, one can restore only one of the following four: 00, 01, 10, 11; in the case of quantum register, one can have the superposition of all four states:

$$|\Psi\rangle = a_0|00\rangle + a_1|01\rangle + a_2|10\rangle + a_3|11\rangle. \qquad (4.17)$$

Now we consider a register composed of L physical bits. Any classical register of that type can store in a given moment of time only one out of 2^L different numbers, i.e. the register can be in only one out of 2^L possible configurations. A quantum register composed of L qubits can store in a given moment of time all 2^L numbers in a quantum superposition:

$$|\Psi\rangle = \sum_{X=0}^{2^L-1} c_X |X\rangle. \tag{4.18}$$

4.8 Quantum Parallelism

Once the register is prepared in a superposition of different numbers we can perform unitary operations on all of them. For example, if qubits are atoms then suitably tuned laser pulses affect atomic states and evolve initial superposition of encoded numbers into different superposition. During such evolution each number in the superposition is affected and as the result we generate a massive parallel computation albeit in one piece of quantum hardware. This is an example of L=4. When we get the unitary transformation as

$$U|0\rangle = \frac{1}{\sqrt{2}}(|0\rangle + |1\rangle), \qquad U|1\rangle = \frac{1}{\sqrt{2}}(|0\rangle - |1\rangle), \tag{4.19}$$

then |0>=|0000> becomes

$$\begin{aligned}
|\Psi\rangle &= U \otimes U \otimes U \otimes U|0000\rangle = U|0\rangle \otimes U|0\rangle \otimes U|0\rangle \otimes U|0\rangle \\
&= \frac{1}{\sqrt{2}}(|0\rangle + |1\rangle) \otimes \frac{1}{\sqrt{2}}(|0\rangle + |1\rangle) \otimes \frac{1}{\sqrt{2}}(|0\rangle + |1\rangle) \otimes \frac{1}{\sqrt{2}}(|0\rangle + |1\rangle) \\
&= \frac{1}{4}(|1111\rangle + |1110\rangle + |1101\rangle + |1100\rangle + |1011\rangle + |1010\rangle + |1001\rangle + |1000\rangle \\
&+ |0111\rangle + |0110\rangle + |0101\rangle + |0100\rangle + |0011\rangle + |0010\rangle + |0001\rangle + |0000\rangle).
\end{aligned} \tag{4.20}$$

This means that a quantum computer can in *only one* computational step perform the same mathematical operation on 2^L different input numbers encoded in coherent superposition of L qubits. In order to accomplish the same task any classical computer has to repeat the same computation 2^L times or one has to use 2^L different processors working in parallel, called **quantum parallelism**. In other words a quantum computer offers an enormous gain in the use of computational resources such as time and memory.

4.9 Quantum Factorization

In order to solve a particular problem computers follow a precise set of instructions that can be mechanically applied to yield the solution to any given instance of the problem. A specification of this set of instructions is called an algorithm.

It is very fast to find the solution of problem as

127 x 129 =? ,

by using known fast algorithms for multiplication. But it is not equally fast to find the solution of the following factorization problem

? x ? = 29083.

Suppose we have a large integer N, which is the product of two prime numbers, the following integer

$$L=[\log_2 N] \tag{4.21}$$

is called the number of digits in binary system.

Skipping details of the computational complexity we only mention that computer scientists have a rigorous way of defining what makes an algorithm efficient or inefficient. For an algorithm to be efficient, the time it takes to execute the algorithm to solve a problem must increase no faster than a polynomial function of the size of the input L and we call the problem as **P** problem. Problems outside class **P** are known as **NP** problems.

The steps it takes to factorize N by classical algorithm on a classical computer are about $2^{L/2}$, thus it is an **NP** problem. We will see that it is a **P** problem by quantum Shor algorithm on a quantum computer.

Quantum computers rely on a different technique to perform efficient factorisation. Indeed it is possible to show that factoring a number can be related to the problem of evaluating the period of a function. To explain how this method works, let us take a simple example and let us imagine we want to find the prime factors of $N = 15$. To do so we pick a random number a smaller than N, for instance $a = 7$, and define a function $f(x) = 7^x \bmod 15$. This function raises 7 to the power of some integer x and takes the remainder of the division by 15. For instance, if $x = 3$, $f(x) = 13$ because $7^3 = 343 = 15\times22 + 13$. Mathematics shows that $f(x)$ is periodic and that its period r can be related to the factors of 15. In our example, we can check easily that $f(x)$ evaluates to 1, 7, 4, 13, 1, 7, 4, 13, 1, 7, 4,... for the values of $x = 0, 1, 2, 3, 4, 5, 6....$ and conclude that the period is $r = 4$. With this information, computing the factors of N only requires to evaluate the greatest common divisor of N and $(a^{r/2} \pm 1)$. In our example computing the greatest common divisor of 15 and $50 = 7^{4/2} + 1$ (or $49 = 7^{4/2} - 1$) returns indeed the values 5 (or 3), the factors of 15.

Obviously classical computers cannot make much of this new method: finding the period of $f(x)$ requires evaluating the function $f(x)$ many times. In fact mathematicians tell us that the average number of evaluation required to find the period is of the same order of the number of divisions needed with the classical algorithm we outlined earlier. With a quantum computer, the situation is completely different: by setting a quantum register in a superposition of states representing 0, 1, 2, 3, 4... it is possible to compute in a single go the values $f(0)$, $f(1)$, $f(2)$ .... These values are encoded in superposed states of a quantum register, retrieving the period from them requires another step (known as a *quantum discrete Fourier transform*), that can also be performed efficiently on a quantum computer.

Now we consider the function that transfers m numbers into n numbers:

$$f: \{0,1,\ldots..2^m-1\}\rightarrow\{0,1,\ldots..2^n-1\}, \tag{4.22}$$

where m and n are positive integers. A quantum computer, because of the unitarity and therefore the irreversibility, has two registers. The first register records the input x as a quantum state $|x\rangle$, and the second register records the output y = f(x) of the first register as the quantum state $|y\rangle$. The different input states are orthogonal with each other, that is, $\langle x|x'\rangle = \delta_{xx'}$; and the different output states are also orthogonal with each other, that is, $\langle y|y'\rangle = \delta_{yy'}$. The unitary evolution operator U acting simultaneously on these two registers determines the evolution of the functions. The operator U acts on the first register $|x\rangle$. The results are recorded into the second register whose state is therefore changed as $|f(x)\rangle$:

$$U|x\rangle|0\rangle = |x\rangle|f(x)\rangle. \tag{4.23}$$

Suppose the first register X has m qubits, then it is able to record any digital number x from 0 to 2^m-1. Expressing each binary number by $a(i)$, then we have:

$$|x\rangle = |a(m-1)\rangle\otimes|a(m-2)\rangle\otimes\ldots\otimes|a(1)\rangle\otimes|a(0)\rangle, \tag{4.24}$$

where $\otimes$ stands for the tensor production. Simply we can omit $\otimes$ and express as:

$$|x\rangle = |a(m-1)\rangle\,|a(m-2)\rangle\ldots|a(1)\rangle\,|a(0)\rangle, \tag{4.25}$$

even as :

$$|x\rangle = |a(m-1)a(m-2)\ldots a(1)a(0)\rangle, \tag{4.26}$$

The relation between the digital number x and the binary numbers $a(i)$ is :

$$x = \sum_{i=0}^{m-1} a(i)2^i, \qquad (a(i) = 0,1) \tag{4.27}$$

For example: $|8\rangle = |1\rangle\,|0\rangle\,|0\rangle\,|0\rangle = |1000\rangle$, $|15\rangle = |1111\rangle$.

Then, the state of the first register is the quantum superposition state of those basic vectors with weights. One time operation of the unitary operator U gets all 2^m values of $f(0),\ldots..f(2^m-1)$. Putting the results into the second register, we have:

$$|\Psi\rangle = U\left(\frac{1}{2^{m/2}}\sum_{x=0}^{2^m-1}|x\rangle\right)|0\rangle \Rightarrow \frac{1}{2^{m/2}}\sum_{X=0}^{2^m-1}|x\rangle|f(x)\rangle. \tag{4.28}$$

For a L-digit register, the single-qubit operator A(j) is that of acting on the j's qubit:

$$A(j)=(1/\sqrt{2})(|0_j\rangle\langle 0_j|+|0_j\rangle\langle 1_j|+|1_j\rangle\langle 0_j|+|1_j\rangle\langle 1_j|). \tag{4.29}$$

When we use the matrix expression, because of

$$|1\rangle = \begin{pmatrix}0\\1\end{pmatrix}, |0\rangle = \begin{pmatrix}1\\0\end{pmatrix}, \langle 1| = (0 \quad 1), \langle 0| = (1 \quad 0), \tag{4.30}$$

Substituting it into A(j) expression, we have

$$A(j) = (1/\sqrt{2})\begin{pmatrix}1 & 1\\1 & -1\end{pmatrix}. \tag{4.31}$$

It is easy to verify that the action of A_j onto the j-qubit, that is :

$$A_j|0_j\rangle=(1/\sqrt{2})(|0_j\rangle+|1_j\rangle);\ A_j|1_j\rangle=(1/\sqrt{2})(|0_j\rangle-|1_j\rangle). \tag{4.32}$$

For a L-digit register, the two-qubit operator B(j, k) is that of acting on the j-qubit and k-qubit:

$$B(j,k)= |0\rangle\langle 0|+|1\rangle\langle 1|+|2\rangle\langle 2|+exp(i\theta_{jk})|3\rangle\langle 3|, \tag{4.33}$$

where we omit the subscripts (j,k), and

$$\begin{aligned}&|0_{jk}\rangle=|0_j\rangle|0_k\rangle=|0_j0_k\rangle, |1_{jk}\rangle=|0_j\rangle|1_k\rangle=|0_j1_k\rangle,\\&|2_{jk}\rangle=|1_j\rangle|0_k\rangle=|1_j0_k\rangle, |3_{jk}\rangle=|1_j\rangle|1_k\rangle=|1_j1_k\rangle;\\&\theta_{jk}=\pi/2^{k-j}.\end{aligned} \tag{4.34}$$

It is easy to verify that the action of B onto the j-qubit and the k-qubit is:

$$B_{jk}|n_{jk}\rangle=|n_{jk}\rangle[\exp(i\pi/2^{k-j})\delta_{n3}], n=0,1,2,3. \tag{4.35}$$

that is, B just shifts the phases θ of the states |3> in the j-qubit and the k-qubit, and keeps the others unchanged.

The core of Shor algorithm (1994) for factoring a large integer N is a quantum method for computing the period of the function

$$f_{a,N}(x) = a^x \bmod N, \tag{4.36}$$

where a is a random integer coprime to N, and $a<N$, $x=0,1,2,3,4,\ldots$. The "mod' function f calculates the remainder after dividing a^x by N. For the definite a and N, we have

$$f(x) = a^x \bmod N. \tag{4.37}$$

Let the period of the function be r, then $f(x) = f(x+r)$, that is,

$$a^x \bmod N = a^{x+r} \bmod N = k, \tag{4.38}$$

which valid for all x, where k is the remainder. We may suppose that

$$a^x = bN + k, a^{x+r} = cN + k. \tag{4.39}$$

where b and $c$are integers, then

$$a^{x+r} - a^x = (c-b)N = 0 \bmod N. \tag{4.40}$$

When $x=0$, and r is an even number, in the condition (if not, we can have another try with different value of a)

$$a^{r/2} \bmod N \neq \pm 1, \tag{4.41}$$

then we have

$$a^r - a^0 = a^r - 1 = (a^{r/2} - 1)(a^{r/2} + 1) = 0 \bmod N. \tag{4.42}$$

This means that we can find

$$C = \gcd(a^{r/2} - 1, N), D = \gcd(a^{r/2} + 1, N), \tag{4.43}$$

where gcd(a, b) is the greatest common divisor of a and b, then we obtain that

$$C \times D = N. \tag{4.44}$$

C and D are two factors of N.

In order to realize the above algorithm, N=15, it is |1111> in the binary system. We have an input-register with 8-digits to record the number N. We have an input-register with 4-digits with the initial state as |0>. The total initial state of two registers is |0>|0>. Then we record x into the input-register,

$$\frac{1}{2^4}\sum_{x=0}^{b}|x\rangle|0\rangle, b = 2^4 - 1. \tag{4.45}$$

By quantum parallel calculation of a unitary mod-function operator, we have

$$\frac{1}{2^4}\sum_{x=0}^{b}|x\rangle|f_{7,15}(x)\rangle. \tag{4.46}$$

where we have select a=7 among {2,4,7,8,11,13,14}; and x can be 0,1,2,3,4,5,6,7,8,...14. Correspondingly, 7^x are 1,7,49,343,2401,16807,...; f=1,7,4,13,1,7,4,13,1,7,.... Obviously, the period of the mod-function f is 4, that is r=4. The state of the two registers is

(1/16)(|0>|1>+|1>|7>+|2>|4>+|3>|13>+|4>|1>+|5>|7>+|6>|4> +|7>|13>+
|8>|1>+|9>|7>+|10>|4>+|11>|13>+|12>|1>+|13>|7> +|14>|4>+|15>|13>), (4.47)

where for simple, in |x>|y>, x in the first sate vector symbol is the input state in digital number system and y in the second sate vector symbol is the output state in digital number system. When we have once measurement on the output register, we will obtain one value among four values of 1,7,4 and 13. Suppose it is 7, then the state of the output register should be |7>, and there is no any other state such as |1>, |4> and |13> (because of the state collapse in quantum mechanics). The state of the two registers is

(1/16) (|1>|7> +|5>|7>+ |9>|7>+ |13>|7>)
=(1/16) (|1>+|5>+|9>+|13>)|7>. (4.48)

The question is how to find the period r. The way to find the period in quantum computer is called discrete Fourier transform.

The discrete Fourier transform is:

$$|x\rangle \Rightarrow \frac{1}{2^{L/2}}\sum_{K=0}^{2^L-1} e^{2\pi i k x/2^L}|k\rangle. \tag{4.49}$$

Here we take L=3 as an example to explain how to realize of the discrete Fourier transform on a quantum computer. Then we will discuss the general situation of L=L to show that it is an efficient algorithm.

The total state of input and output registers is the superposition state of $|x>|f(x)> = |x, f(x)>$:

$$|\Psi\rangle = \sum_{k,n} c_{k,n} \,|\, k, f(n) >, \tag{4.50}$$

where *k and n* are all integers. In the example considered, it becomes:

$$|\Psi\rangle = \frac{1}{\sqrt{8}} \sum_{x} |\, x, f(x) >, \tag{4.51}$$

that is,

$$|\Psi>=(1/\sqrt{8})(\,|0,f(0)>+\,|1,f(1)>+\,|2,f(2)>+\,|3,f(3)>+ \\ |4,f(4)>+\,|5,f(5)>+\,|6,f(6)>+\,|7,f(7)>). \tag{4.52}$$

Now we make the discrete Fourier transform on the input register,

$$|x\rangle \Rightarrow \frac{1}{2^{L/2}} \sum_{K=0}^{2^L-1} e^{2\pi ikx/2^L} |k\rangle. \tag{4.53}$$

For the case considered, the above equation becomes

$$|x\rangle \Rightarrow \frac{1}{\sqrt{8}} \sum_{K=0}^{7} e^{2\pi ikx/8} |k\rangle. \tag{4.54}$$

The total state of input and output registers is the summation of 64 terms:

$$|\Psi\rangle \Rightarrow |\Psi'\rangle = \frac{1}{8} \sum_{k,x=0}^{7} e^{2\pi ikx/8} \,|\, k, f(x) >. \tag{4.55}$$

Because the "mod' function

$$f_{a,N}(x) = a^x \bmod N, \tag{4.56}$$

is a period function with period r, and further suppose that r is a even number (otherwise we can try to select another a until find an even r). Now we do not know the value of r and it is just what we are going to find. In the following, we suppose that $r = 2$, then $f(0)=f(2)=f(4)=f(6)$, $f(1)=f(3)=f(5)=f(7)$, above summation of 64 terms becomes the summation of 58 terms (the phrase factors of the states of k=0 are all equal to 1, so 8 terms are combined into two terms, and the front coefficient becomes 1/2). The number of states left is 16. The state of input and output is as following according to the sequence of input states:

$$
\begin{aligned}
|\Psi'\rangle = & \ (1/2)|0\rangle\{|f(0)\rangle+|f(1)\rangle\}+ \\
& (1/8)|1\rangle\{|f(0)\rangle(1+e^{i\pi/2}+e^{i\pi}+e^{i3\pi/2})+|f(1)\rangle(e^{i\pi/4}+e^{i3\pi/4}+e^{i5\pi/4}+e^{i7\pi/4})\}+ \\
& (1/8)|2\rangle\{|f(0)\rangle(1+e^{i\pi}+e^{i2\pi}+e^{i3\pi})+|f(1)\rangle(e^{i\pi/2}+e^{i3\pi/2}+e^{i5\pi/2}+e^{i7\pi/2})\}+ \\
& (1/8)|3\rangle\{|f(0)\rangle(1+e^{i3\pi/2}+e^{i3\pi}+e^{i9\pi/2})+|f(1)\rangle(e^{i3\pi/4}+e^{i9\pi/4}+e^{i15\pi/4}+e^{i21\pi/4})\}+ \\
& (1/8)|4\rangle\{|f(0)\rangle(1+e^{i2\pi}+e^{i4\pi}+e^{i6\pi})+|f(1)\rangle(e^{i\pi}+e^{i3\pi}+e^{i5\pi}+e^{i7\pi})\}+ \\
& (1/8)|5\rangle\{|f(0)\rangle(1+e^{i5\pi/2}+e^{i5\pi}+e^{i15\pi/2})+|f(1)\rangle(e^{i5\pi/4}+e^{i15\pi/4}+e^{i25\pi/4}+e^{i35\pi/4})\}+ \\
& (1/8)|6\rangle\{|f(0)\rangle(1+e^{i3\pi}+e^{i6\pi}+e^{i9\pi})+|f(1)\rangle(e^{i3\pi/2}+e^{i9\pi}+e^{i15\pi/2}+e^{i21\pi/2})\}+ \\
& (1/8)|7\rangle\{|f(0)\rangle(1+e^{i7\pi/2}+e^{i7\pi}+e^{i21\pi/2})+|f(1)\rangle(e^{i7\pi/4}+e^{i21\pi/4}+e^{i35\pi/4}+e^{i49\pi/4})\}.
\end{aligned}
\tag{4.57}
$$

In above, because of the action of the phase factors, all terms are canceled out except the first and fourth term. At last, there are four terms left as following:

$$|\Psi'\rangle = (1/2)\{|0, f(0)\rangle+|0, f(1)\rangle+|4, f(0)\rangle+e^{i\pi}|4, f(1)\rangle\}. \tag{4.58}$$

Now we see that because of the periodicity of "mod' function and the coherence of quantum mechanics, the last result is greatly simplified.

In this example, measuring the states of the input register, we will get states |0> and |4> with probability 1/2 respectively. According to Shor algorithm,

$$k=0, \ d/r, \ 2d/r, \ 3d/r, \ \ldots,(r-1)d/r, \tag{4.59}$$

where d is the possible number of states of input register. In the example, d=8. We have measured that k=0 or 4. From above, we get the period r=2.

The discrete Fourier transformation is realized by following quantum operations, that is, by using the single-qubit operator A(j) and the two-qubit operator B(j,k) sequentially:

$$|2\rangle \Rightarrow A(0)\ B(0,1)\ B(0,2)\ A(1)\ B(1,2)\ A(2)\ |2\rangle. \tag{4.60}$$

Now we calculate it from right to left sequentially as following:

$A(2)|2\rangle = A(2)|010\rangle$ (A acts on the second qubit from right in the order of 0, 1, 2)
$= (1/\sqrt{2})(|0\rangle+|1\rangle)|1\rangle|0\rangle$,

$$=(1/\sqrt{2})|010\rangle+|110\rangle. \qquad (4.61)$$

$$B(1,2)A(2)|2\rangle=B(1,2)(1/\sqrt{2})(|010\rangle+|110\rangle)$$
$$=(1/\sqrt{2})[|010\rangle+i|110\rangle]. \qquad (4.62)$$

(B shifts the phrases of the first and second qubit from right in the order of 0,1,2)

$$A(1)\,B(1,2)\,A(2)\,|2\rangle=A(1)(1/\sqrt{2})[\,|010\rangle+i|110\rangle]$$
$$=(1/2)[|000\rangle-|010\rangle+i(|100\rangle-|110\rangle)], \qquad (4.63)$$

$$B(0,1)\,B(0,2)\,A(1)\,B(1,2)\,A(2)\,|2\rangle=$$
$$=(1/2)[|000\rangle-|010\rangle+i(|100\rangle-|110\rangle)],\text{(unchanged)}. \qquad (4.64)$$

The last step of the discrete Fourier transform reaches the result:

$$A(0)\,B(0,1)\,B(0,2)\,A(1)\,B(1,2)\,A(2)\,|2\rangle=$$
$$=A(0)(1/2)[|000\rangle-|010\rangle+i(|100\rangle-|110\rangle)]$$
$$=(1/\sqrt{8})[|000\rangle+|001\rangle-|010\rangle-|011\rangle+i(|100\rangle+|101\rangle-|110\rangle-|111\rangle)]. \qquad (4.65)$$

Turning over all the qubits obtained from the discrete Fourier transform, $(ijk)\Rightarrow(kji)$, we have

$$(1/\sqrt{8})[|000\rangle+|100\rangle-|010\rangle-|110\rangle+i(|001\rangle+|101\rangle-|011\rangle-|111\rangle)]$$
$$=(1/\sqrt{8})[|0\rangle+|4\rangle-|2\rangle-|6\rangle+i(|1\rangle+|3\rangle-|5\rangle-|7\rangle)]. \qquad (4.66)$$

This is complete coincide with the result of transformation formula of the discrete Fourier transform:

$$|2\rangle\Rightarrow\frac{1}{\sqrt{8}}\sum_{K=0}^{7}e^{\pi ik/2}|k\rangle$$
$$=\frac{1}{\sqrt{8}}(|0\rangle-|2\rangle+|4\rangle-|6\rangle+i(|1\rangle-|3\rangle+|5\rangle-|7\rangle)) \qquad (4.67)$$

In the general case of L=L, the discrete Fourier transform can be realized by following procedure:

A(L-1)|x>,
A(L-2)B(L-2,L-1)[A(L-1)|x>],(that in [ ] is the result obtained by the first step)
A(L-3)B(L-3,L-2)B(L-3,L-1)[A(L-2)B(L-2,L-1) A(L-1)|x>],
(that in [ ] is the result obtained by the second step)
A(L-4)B(L-4,L-3)B(L-4,L-2)B(L-4,L-1)[the result obtained by the third step]

And so on, the expression of the general term is:

$$[A(0)B(0,1)B(0,2)\ldots B(0,L-1)]$$

$$
\begin{aligned}
&[A(1)B(1,2)B(1,3)\ldots B(1,L-1)] \\
&[A(2)B(2,3)B(2,4)\ldots B(2,L-1)] \\
&[A(3)B(3,4)B(3,5)\ldots B(3,L-1)] \\
&\ldots \\
&[A(L-4)B(L-4,L-3)B(L-4,L-2)B(L-4,L-1)] \\
&[A(L-3)B(L-3,L-2)B(L-3,L-1)] \\
&[A(L-2)B(L-2,L-1)] \\
&[A(L-1)]|x\rangle]. \qquad (4.68)
\end{aligned}
$$

For the case of L=3,

$$|2\rangle \Rightarrow [A(0)B(0,1)B(0,2)][A(1)B(1,2)]A(2)|010\rangle, \qquad (4.69)$$
$$|4\rangle \Rightarrow [A(0)B(0,1)B(0,2)][A(1)B(1,2)]A(2)|100\rangle. \qquad (4.70)$$

L=4,

$$|5\rangle \Rightarrow [A(0)B(0,1)B(0,2)B(0,3)][A(1)B(1,2)B(1,3)] [A(2)B(2,3)]A(3)|0101\rangle, \qquad (4.71)$$
$$|9\rangle \Rightarrow [A(0)B(0,1)B(0,2)B(0,3)][A(1)B(1,2)B(1,3)] [A(2)B(2,3)]A(3)|1001\rangle. \qquad (4.72)$$

From above, we can see that the discrete Fourier transform wit L-qubits needs L times A (j) operations, L(L-1)/2 times B(j, k) operations, and the total number of operations is L(L+1)/2, therefore the discrete Fourier transform is efficient.

Now we give a brief comparison the quantum factorization with the classical one. A classical algorithm to factor an integer number N on a classical computer is based on checking the remainder of the division of N by some number p smaller than square root of N. If expressed in binary system, $N = [2^L]$, $\sqrt{N} \approx 2^{L/2}$. Let L=200, $2^{L/2}=2^{100}\approx 10^{30}$, if the remainder is 0, we conclude that p is a factor. This method is in fact very inefficient: with a computer that can test for 10^{12} different p's per second (this is faster than any computer ever built), the average time to find the factor of a 200-digit long number would be of the order of 10^{18} seconds, which exceeds the age of the universe 10^{17} seconds!

But to find the factor on a quantum computer with Shor algorithm, because L (L+1)/2 $\leq L^2$, when L>5, $2^L > L^2$, $L^2 \approx 4\times 10^4$, with the same rate of calculation , it takes about 10^{-8} seconds!

5. Conclusion

As a conclusion, we have confidence to think that the recently theoretical and experimental developments on entanglement and nonlocality are solid and important for our understanding the implication of quantum mechanics and that the developments on quantum computers where quantum physics, information theory and computer science mash heads in hopes of better computers and cleaner understandings of physics, are very important for the entering the new age of quantum information, but further developments will not be easy, therefore will be wonderful and attractive.

Acknowledgements

Z.J.Z. acknowledges support from National Science Foundation of China Grant No. 69773052.

References

1. R. Feynman, *Int. J. Theor. Phys.* **21** (1981) 486. *Int. J. Theor. Phys.* **21** (1982), 467. *Opt. News*, **11** (1982), 11.
2. A. Einstein, B. Podolsky and N. Rosen, Can quantum-mechanical description of physical reality be considered complete? *Phys. Rev.*, **47** (1935), 770-780.
3. N. Bohr, Can quantum-mechanical description of physical reality be considered complete? *Phys. Rev.*, **48** (1935), 696-702.
4. J. S. Bell, *Speakable and Unspeakable in Quantum Mechanics* (Cambridge Univ. Press, Cambridge, England, 1988
5. A. Aspect, Proposed experiment to test the nonseparability of quantum mechanics, Phys. Rev., D**14** (1976), 1944-51.
   A. Aspect, P. Grangier, and G. Roger, *Phys. Rev. Lett.*, **47** (1981),460.
   A. Aspect, P. Grangier, and G. Roger, Experimental Realization of Einstein-Podolsky-Rosen-Bohm Gedankenexperiment: A New Violation of Bell's Inequalities, *Phys. Rev. Lett.*, **49** (1982), 91.
   A. Aspect, J. Dalibard and G. Roger, Experimental Test of Bell's Inequalities Using Time-Varying Analyzers, *Phys. Rev. Lett.*, **49** (1982), 1804.
6. Greenberger D. M., M. A. Horne, & A. Zeilinger, in *Bell's Theorem, Quantum Theory, and Conceptions of the Universe*, edited by M. Kafatos, Kluwer Acedemics, Dordrecht, The Netherlands, 73(1989).
   Greenberger D. M., M. A. Horne, A. Shimony, & A. Zeilinger, *Am. J. Phys.* **58**, 1131 (1990).
   Kwiat P., Mattle K., Weinfurter H., Zeilinger A., Sergienko A. & Shih Y., New High-Intensity Source of Polarization-Entangled Photon Pairs, *Phys. Rev. Lett.* **75**, 4337,(1995)
   Bouwmeester D, Jian-Wei Pan, Mattle K., Eibl M., Weinfurter H, and Zeilinger A., Experimental Quantum Telepotation, *Nature*, **390** (1997), 575-579.
   Pan J.-W., Bouwmeester D, Eibl M., Weinfurter H, and Zeilinger A., Experimental Entanglement Swapping: Entangling Photons That Never Interacted, *Phy.Rev.Lett.***80** (1998), 3891-3894
   Pan J.-W., M. Daniell, H. Weinfurter, & A. Zeilinger, Phys. Rev. Lett. **82** (1999), 1345.
   Brassard G. et al., Cost of Exactly Simulating Quantum Entanglement with Classical

Communication, *Phys. Rev. Lett.* **83**, 1874-1877 (1999)
Pan J.-W. et al., Experimental test of quantum nonlocality in three-photon GHZ entanglement, (submitted, to be appeared in *Nature*, 1999,8)

7. Landauer R., *IBM J. Res. Dev.*, **5**, 183(1961).
Bennett C. H., *IBM J. Res. Dev.*, **17** (1973), 525
Benioff P., *J. Stat. Phys.* 22, 563(1980)
Deutsch D., *Proc. R. Soc. London*, A **400**, 97 (1985).
Bennett, C. H., *Phys. Rev. Lett.*, **68**, 3121(1992)
Bennett, C. H., Brassard G., Crepeau C., Jozsa R., Peres A. and Wootters W. K., Teleporting an unknown quantum state via dual classic and Einstein-Podolsky-Rosen channels, *Phys. Rev. Lett.*, **70**, 1895(1993)
Barenco A., Quantum Physics and Computers, in *Contemporary Physics*, **37** (1995), 375-389.
Barenco, A., Deutsch D. and Ekert A., Conditional Quantum Dynamics and Logic Gates, *Phys. Rev. Lett.*, **74** (1995), 4083,

8. Shor P. W., in *Proceedings of the 35th Annual Symposium on the Foundations of Computer Science*, edited by S. Goldwasser (IEEE Computer Society Press, Los Alamitos, CA), 124 (1994).
Ekert A. and R. Jozsa, Quantum Computation and Shor's Factoring Algorithm, *Review of Modern Physics*, **68**, 733-753, (1996).

9. Gershenfeld N. A. and Chuang I. L., Bulk Spin-Resonance Quantum Computation, *Science*, **277**, 1689(1997)
Hakamura,Y.,YU.A. Pashkin & J.S. Tsai, Coherent control of macroscopic quantum states in a single-Cooper-pair box, *Nature*, **398**, 786-788, (1999)
Jonathan D. and Plenio M. B., Minimal Conditions for Local Pure-State Entanglement Manipulation, *Phy.Rev.Lett.***83**, 1455-1458 (1999)
Lo H. and Popescu S., Classical Communication Cost of Entanglement Manipulation: Is Entanglement an Interconvertible Resource? *Phys.Rev.Lett.* **83**, 1459(1999)
Preskill,J., and Alexei Kiteev, *Quantum Information and Computation*, 1999, www.theory.caltech.edu/~preskill/ph229
Quiroga L. and Johnson F.N., Entangled Bell and Greenberger-Horne-Zeilinger States of Excitons in Coupled Quantum Dots, *Phys. Rev. Lett.* **83**, 2270(1999)
Sctulz M., The end of the road for silicon, *Nature*, **399** (1999) 729.
Vidal G., Entanglement of Pure States for a Single Copy, *Phys. Rev. Lett.* **83**, 1046-1049 (1999)

SUBJECT INDEX

www.ingramcontent.com/pod-product-compliance
Lightning Source LLC
Chambersburg PA
CBHW050638190326
41458CB00008B/2323